LABORATORY ANIMAL HANDBOOKS No. 12

PARASITES OF LABORATORY ANIMALS

By

DAWN G OWEN

1992
LONDON

Published for Laboratory Animals Ltd
by Royal Society of Medicine Services Limited

Published in the United Kingdom for Laboratory Animals Ltd
by Royal Society of Medicine Services Limited,
1 Wimpole Street, London W1M 8AE.

British Library Cataloguing in Publication Data
Owen, D. G. (Dawn G)
Parasites of laboratory animals.
I. Title
591.5249

ISBN 1-85315-159-9

Cover drawings (clockwise from top left): *Myobia musculi*, adult female; *Heligmosomoides (Nematospiroides) dubiuus*, adult male; *Gliricola porcelli*, adult female; *Syphacia obvelata*, adult female (drawings by Gary Moores). Oocyst of *Eimeria intestinalis* (reproduced from Catchpole & Norton 1979, *Parasitology* **79**, 252, with kind permission of the authors and Cambridge University Press)

Typeset by Dobbie Typesetting Limited, Tavistock, Devon.
Printed in Great Britain by Henry Ling Ltd, at the Dorset Press, Dorchester, Dorset

To George Porter

TO A LOUSE

ON SEEING ONE ON A LADY'S BONNET AT CHURCH

HA! whare ye gaun, ye crowlin ferlie!
Your impudence protects you sairly:
I canna say but ye strunt rarely,
Owre gauze and lace;
Tho', faith, I fear ye dine but sparely
On sic a place.

Ye ugly, creepin', blastit wonner,
Detested, shunn'd, by saunt an' sinner,
How dare ye set your fit upon her,
Sae fine a lady!
Gae somewhere else, and seek your dinner
On some poor body.

O wad some Pow'r the giftie gie us
To see oursels as ithers see us!
It wad frae mony a blunder free us,
An' foolish notion!
What airs in dress an' gait wad lea'e us,
An' ev'n devotion!

Robert Burns (1759–96)

CONTENTS

ACKNOWLEDGEMENTS

Much of the material in this book was accumulated and collated during the long period that I worked in the Pathology Department of the Medical Research Council Laboratory Animals Centre at Carshalton. I would like to take this opportunity to thank my many colleagues and associates of that time for their help and encouragement, and also our director of the time, Dr J Bleby, now of the Royal Veterinary College, for making it possible. I also thank Dr Margery Wood of the Chemical and Biological Defence Establishment, Porton Down, for her appraisal and heroic editing of earlier drafts of the book; Dr David Dunne of the Cambridge University Pathology Department for his advice and comments on the manuscript; and Mr Christopher Norton of the Central Veterinary Laboratory who has allowed the reproduction of his beautiful drawings of coccidial oocysts (see Fig. 3.11). There are many others too numerous to mention who provided specimens or prepared histological material, but a special mention must be made of the various sections at the British Museum of Natural History who have given so many references and so much help and advice over the years.

The following illustrations, first published in my previous book *Common Parasites of Laboratory Rodents and Lagomorphs* (1972; Medical Research Council Laboratory Animals Centre Handbook No. 1), are reproduced here with the kind permission of the Controller of Her Majesty's Stationery Office: *Ectoparasites*—Figs. 2.1, 2.2, 2.3A, 2.7B, 2.8B, 2.9B, 2.10BC, 2.12A, 2.17A, 2.19A; *Endoparasites*—Figs. 3.3, 3.23AB, 3.26AB, 3.37ABCD, 3.38ABD, 3.39AB, 3.42ABC, 3.43AB, 3.55AC.

Finally, I am most grateful to Mr Robert Legg, who took many of the photographs; Miss Lynda Norris, who patiently typed and sorted the manuscript at various stages of its metamorphosis; and Delia Siedle, of Royal Society of Medicine Services, who has gathered it all together into its final form.

Dawn G Owen

INTRODUCTION

The main object of this handbook is to present an account, based on personal experience, of the common parasites of a range of laboratory-bred mammals, principally rodents, together with those of some species of wild monkey regularly used for research.

Because of the improvement in standards of production, husbandry and housing, the majority of laboratory-bred rats and mice are now free from a range of parasites found in their wild counterparts. This handbook contains specimens collected during post-mortem examination or received for identification from breeders or users of laboratory rodents and lagomorphs over a period of many years. Until the breeding of primates in captivity can be established on a scale large enough to supply the demands of research, some work will continue to be carried out on wild-caught animals. These will usually contain, despite quarantine periods and drug treatment, a range of specific parasites that might well be of significance to the research worker—hence the inclusion of so many exotic parasites specific for this group. Because of the enormous range of both parasite species and hosts, only the most commonly used species of Old and New World monkeys are considered in detail. Other species are noted whenever it is felt this would be useful.

The text is not intended as a definitive work but as a practical aid to research workers who may not necessarily be interested in parasites as such but who are likely to come across them during the course of their investigations. The animals considered here include rodents—rats (*Rattus norvegicus*), mice (*Mus musculus*), hamsters (*Mesocricetus auratus*), gerbils (*Gerbillus gerbillus* and *Meriones unguiculatus*) and guinea pigs (*Cavia porcellus*); the rabbit (*Oryctolagus cuniculus*); and six commonly used species of primates—rhesus and cynomolgus monkeys (*Macaca mulatta* and *M. fascicularis*) and baboons (mainly *Papio dogeura*) from the Old World, and squirrel monkeys (*Saimiri sciureus*), marmosets (*Callithrix jacchus*) and tamarins (*Saguinus oedipus*) from the New World.

Each group of parasites is described with special reference to those of particular interest or pathogenicity and, where necessary, photographs show details of the anatomy and pathology to aid diagnosis. The host-parasite lists (Appendix A) include comments on the possible clinical effects of the organism and further details can be found in the list of references. Treatment of some pathogenic parasites may be essential and the final section gives a brief summary of control methods and some drugs described in the literature as having been successfully used in rodent colonies.

SECTION 1

TECHNIQUES AND SERODIAGNOSIS

TECHNIQUES

COLLECTION AND HANDLING OF BIOLOGICAL MATERIAL

Care should always be taken in the handling of all biological material. A sound, routine procedure should be established whenever animal tissues are to be regularly examined. Suitable laboratory coats and gloves should be worn, particularly during post-mortem procedures.

Samples of blood or faeces should be taken into sterile containers together with the medium required. When sending any samples by post it is important to ensure that the sample is in a suitable screw-top plastic container, sealed with waterproof tape and placed in a plastic bag within a padded box so that the container is not broken on the journey.

It is of particular importance when dealing with wild species, especially monkeys, however long they might have been in quarantine or in the primate house, to treat their tissues with as much caution as you would the live animal. Over the years there have been several serious instances of monkeys passing on often fatal virus infections to scientists and animal handlers (Hanson *et al.* 1967). In Germany, 30 people became ill and seven died as a result of handling sick African green monkeys (*Cercopithecus aethiops*) which were carrying an agent now known as Marburg virus (Smith *et al.* 1967). *Herpes virus simiae* (B-virus) is another pathogen which is known to be communicated by aerosol as well as scratch and bite and which has been the cause of high mortality in infected monkey handlers (Davidson & Hummeler 1960; Brack 1987).

There is a wide variety of bacteria transmissible to man from monkeys, e.g. *Mycobacterium tuberculosis*, *Salmonella* sp. and *Shigella* sp., and some parasites such as *Entamoeba histolytica* and *Giardia* sp. are possible zoonotic agents (Neal 1969). Whenever handling monkeys, suitable coveralls, boots, gloves, hats and face-masks should be worn. Clean protective clothing should be put on before entering the unit and all items removed before leaving. All the time and effort is wasted if staff wander around outside the monkey house in the boots and coveralls used inside!

Biological material, such as blood films, swabs or faecal samples, should be brought out of the area in air-tight containers, disinfected on the outside, and taken directly to the laboratory. Where possible, material should be fixed before removal from the primate quarters. All utensils and protective clothing must be sterilized after use and all waste material incinerated.

EXAMINATION OF THE LIVING ANIMAL

Unless the animal is restrained or sedated, a thorough physical examination is difficult without assistance. In the case of the wild monkey, investigations often have to be limited to the collection and examination of stool samples if it cannot be sedated. However, a great deal of information can be gained from faecal samples if sampling is carried out properly and at regular intervals. Wet preparations and saturated salt flotations (*see* Appendix B) will give a reasonably accurate picture of the parasitic burden. Clean trays should be used for collection and either filled with clean sawdust or, preferably, lined with clean paper. The sample must be examined while fresh and if an accurate picture of the parasitic burden is to be gained, several such samples taken over a period of approximately two to three weeks must be examined.

The sedated animal may be given a more thorough examination. Skin and fur may be investigated for the presence of dermatophytic fungi or ectoparasites such as ticks or lice. Blood samples can be taken for smears and other tests, for example antibody detection, and skin scrapings taken of scurfy areas. Where no fresh stool sample is available, rectal swabs can also be taken. This is an opportunity to take nose and throat swabs to look for bacterial infections and, in the case of macaques, to look for *Anatrichosoma* sp.; this nematode burrows into the nasal epithelium, and its ova are not always found in the faeces.

BLOOD SAMPLING

Collection of blood in sufficient amounts for serum production or larval counting is usually carried out using a syringe and needle and sampling from a major blood vessel. In the rabbit the peripheral ear vein is used. The site is carefully shaved and swabbed with 70% ethanol before extracting the sample. The blood may also be obtained by puncturing the vessel with a sterile hypodermic needle and allowing it to drip into the sample vessel. Larger quantities can be obtained by heart puncture, which must be done under anaesthesia and involves a risk to the animal. Collection of blood samples from some species such as the guinea pig is often difficult. Small amounts may be obtained by puncture of the marginal ear vein or, under anaesthesia, by syringe from the saphenous vein at a point above the tarsal joint. Heart puncture introduces the risk of haemorrhage into the pericardium (Poole 1972). In rodents small samples may be obtained by clipping the skin at the extreme tip of the tail. This is particularly effective if the animals are kept in a warm room or under a warm lamp prior to bleeding. Enough blood can be collected from a rat and from a good size mouse for microbiological techniques and to do a PCV (haematocrit) estimation as well as to make blood films. Small samples can also be collected by syringe and needle from the tail vein, but larger volumes should only be obtained under anaesthetic either directly from the heart or from the jugular vein (Kassel & Levitan 1953). Blood from primates is best collected by syringe from the femoral plexus or from the saphenous vein at a point behind and below the knee.

Samples of 50 μl can be collected in disposable capillary pipettes (Accupette pipettes, Dade Diagnostic Inc., American Hospital Supply Corporation, 185 Delaware Parkway, Miami, Florida 33152, USA) and can be used in large-scale screening where it is not possible to collect sufficient whole blood for normal serology (Ollett 1951). The drop of blood is sucked into the pipette and ejected on to a filter paper. The dried blood spot is eluted overnight in 0.4 ml distilled water; the filter paper is removed and the eluted sample used as a final dilution of 1 in 16. This method was successfully used by Gannon (1978) in a survey of *Encephalitozoon cuniculi* in laboratory rodents and lagomorphs. The filter papers with dried blood spots can be sealed in polythene bags and will remain in good condition for several months if stored at −20 °C. The most important point to note is the need to keep the samples dry and the use of silica gel is recommended.

POST-MORTEM EXAMINATION

External examination: surface-dwelling parasites

This should consist of a thorough general examination taking into account all signs of possible parasitic infection such as a rough or patchy coat, loss of fur or soiled hindquarters. At this point note should be taken of the animal's history in order to rule out lesions due to accident or fighting.

The majority of fur-dwelling arthropod parasites are strictly host-specific, and many have preferred locations on the host for feeding and for egg-laying. This is the case with many lice and mites that live on the body surface, such as the chewing lice of guinea pigs, *Gyropus ovalis* and *Gliricola porcelli*, and the mange-producing mites of rodents, *Myobia musculi* and *Myocoptes musculinus*. All these surface-dwelling parasites can usually be located using a binocular dissecting microscope with a magnification of ×20. The microscope should preferably have an adjustable swivel head on a heavy base without a fixed light in order to accommodate the larger species of animal to be examined such as the rabbit. The fur should be parted and the skin and hair systematically searched using a pointed needle, especially in the regions preferred by parasites. Any parasite found should be picked up using a needle or a fine brush moistened with 70% ethanol, and transferred either to 70% ethanol or to a drop of Berlese fluid (chloral gum) (BDH Chemicals, Freshwater Road, Dagenham, Essex, UK) on a clean slide. Identification of the parasite should then be made on the basis of its morphology and also by consideration of the host species and anatomical location where it was found.

Even if the adult arthropods have disappeared from the host's body because of cooling or as a result of treatment, the ova will remain and can still make identification possible. Eggs can be plucked out and mounted in chloral gum; those of mites and lice are distinctive and usually easily distinguishable. Searching the skin of small rodents is easily accommodated under the dissecting microscope and a general search of the body can be attempted. Larger animals are more difficult and it is necessary to know the preferred site for egg-laying in order to reduce the search time. For example, of the two common chewing lice of guinea pigs, *Gliricola porcelli* lays its eggs on small fine hairs, usually most easily found on the back legs around the anus and rump. *Gyropus ovalis*, however, prefers the head and face, and eggs can easily be found around the edge of a bare area of skin at the back of the ears. Lice in non-human primates are restricted to a species of chewing louse, *Trichodectes* sp. (Stiles & Hassall 1929), and some blood-sucking species, *Pediculosus* and *Pedicinus* (Hopkins 1949; Fiennes 1972), all of which may be found anywhere on the body surface but rarely in large numbers and mainly in the great apes. Neither group of lice has been implicated as a vector of other infections nor as a cause of overt disease in primates.

In wild rabbits, *Spilopsyllus cuniculi* is a common flea; the adults are found in the fur (Mead-Briggs 1965) and show a predilection for the ears. The mite *Myocoptes musculinus* in the mouse tends to prefer the area around the shoulders and flanks. *Myobia musculi*, on the other hand, can usually be found on the skin around the face and head. *Psoroptes cuniculi*, the ear canker mite of the rabbit, can be seen early in the infection deep in the external aural canal by means of an auriscope. At post-mortem examination, slitting down the side of the ear canal will reveal the parasite. Clinical signs become apparent as the auditory meatus fills with exudate (Sweatman 1958). The lesions can extend to the edge of the pinna and, in some cases, on to the face. Other mites of the rabbit, such as *Cheyletiella parasitivorax*, are found generally on the body surface. Searching the finer hair of the head, neck and face should reveal them if they are present. Active mites are best observed in the weanling rabbit (Beesley 1963; Kraus 1974).

Many fur-dwelling species will be found during a general inspection. *Chirodiscoides caviae* in the guinea pig, and *Listrophorus gibbus* in the rabbit, are prolific, non-pathogenic inhabitants of the fur where they may be found clinging half way up the hair shaft. It is here that they cement their long, dark brown eggs.

External examination: skin lesions

The first signs of mange are usually coarsening of the coat, and irritation followed by thickening of the skin as a result of continual scratching which eventually causes lesions that may become secondarily infected. At this stage the parasites may be fairly sparse in the affected area, as the thin coat and inflamed skin are an unstable environment. The search for surface-dwelling parasites should therefore be made in the area around the lesions.

Internal examination

In the majority of host species it is worthwhile examining the under-surface of the pelt. Peel back as much as possible of the skin over the whole dorsal and ventral surfaces. In mice there is the chance that this may be rewarded by the rare finding of the small white nodules of *Psorergates simplex*, a follicle-inhabiting mite whose 'nests' can be seen on the under-surface of the skin. *Sarcocystis* sp. could be observed at this point, if there were a heavy infection, as fine white stripes in the musculature. A portion of muscle from either the thigh, scapular or lumbar regions should be examined histologically for the presence of this parasite. Where the nematode *Trichinella spiralis* is suspected in the muscles, a piece of diaphragm should be taken and compressed between two slides. In heavy infections the larvae are readily seen at this site. To detect lighter infections, a muscle-digestion technique should be used (*see* Appendix B). Filarial worms may be found coiled in the connective tissues of many New World primates. In Old World monkeys larval cestodes such as *Taenia solium* (*Cysticercus cellulosae*) have been described as readily discernible in the muscles.

In the body cavities of New World primates, care should be taken to look for filarial worms which may be found coiled freely in the thorax or abdomen. Other parasites which might be noticed at this stage are the larvae of pentastomids within the mesentery, either as distinctive coils or as nodules, and cestode larvae such as *Multiceps* sp. and *Echinococcus granulosus*, which produce sizeable cysts in species such as *Macaca mulatta* but are only rarely reported.

Many parasites produce characteristic macroscopic changes in the organs they inhabit. For example, *Capillaria hepatica* produces an unmistakable white fibrous lesion crossing the surface of the liver capsule in a wide range of wild relatives of laboratory rodents. The larval cyst of the cat tapeworm, *Taenia taeniaeformis* (*Cysticercus fasciolaris*), is seen as a large pearly-white body usually protruding conspicuously from the ventral surface of the liver capsule; usually only one cyst is present but sometimes more. In rabbits the cysts of dog tapeworms, for example *Taenia pisiformis* (*Cysticercus pisiformis*), are commonly found loosely attached in the mesentery (Bull 1953); these small oval-shaped cysts are usually present in considerable numbers and are easily recognized.

In rabbits, the coccidian *Eimeria stiedae* can produce extensive gross liver lesions, e.g. enlargement and thickening of the bile ducts; also the microsporidian *Encephalitozoon* (*Nosema*) *cuniculi* can cause extensive necrosis, seen as pitting of the kidney surface which is pale in colour. A similar phenomenon in the kidneys is found in guinea pigs infected with *Klossiella caviae*.

The fibrous nests of *Pneumonyssus* sp. in the lungs of Old World monkeys can be the size of a small pea and easily palpated; and in the large intestine the larvae of *Oesophagostomum* sp. produce a thickening of the mucosa, seen as creamy or brownish nodules, the size of a small pea, through the serosa.

Many other parasites are not so easily located without deeper examination. Small flukes in the bile duct may result in an unspecific swelling and congestion of the liver, or might pass unnoticed depending on the stage and level of infection.

Even in the absence of gross clinical lesions, a careful examination should be made of all the organs and samples taken for histology. Each organ should be removed separately and given a thorough inspection; for example, lungs should be examined for filarial worms or mites, bronchi should be split open as far as possible and swabbed for the presence of larvae. Bile should be collected, particularly in the case of rabbits for wet smears, bile ducts should be opened, as far as possible, and the liver cut into thin slices. Nodules in the mesentery should be carefully dissected and the pancreas teased out in saline and examined microscopically and, where possible, samples taken for histological examination.

The entire alimentary tract should be removed and different regions placed in separate receptacles containing Normal saline (0.85% NaCl). Samples should be taken for histology before further examination. Each region should then be slit open with gut scissors in a dish containing a little saline and the incised edges laid flat. The exposed inner surface should be washed gently with saline and examined microscopically. Any worms or flukes adhering to the surface can then be removed to the correct media. Often it is necessary to examine the surface closely and take careful scrapings using a dissecting microscope to ensure that small flukes or fine nematodes such as *Capillaria* or *Gongylonema* sp. are not missed.

Samples of the gut contents from each region should be examined in a petri dish against a dark background, and wet preparations of the contents examined microscopically for the presence of moving parasites. Smears of gut contents and blood films should be fixed at this time. When dealing with the smaller rodents the whole of the small intestine can be placed in one dish and the contents expressed by holding the duodenal portion with forceps and squeezing out the contents using a blunt seeker. This method causes adult cestodes to release their scolices from the mucosa. Finally, a sample of the rectal contents should be taken and homogenized in 10% formol saline. It is important that the sample is well broken up when the fixative is added. The suspension can then be filtered and treated by either flotation or sedimentation techniques (*see* Appendix B).

PREPARATION OF MATERIAL FOR IDENTIFICATION

Ectoparasites

Skin scrapings are made using a sterile scalpel held at an angle of 45° and scraped along the surface of the skin at the edge of the lesion until the upper layers of the epidermis are removed. The small skin samples should then be placed with a drop of 10% sodium hydroxide on a slide, warmed carefully over a flame until the material has disintegrated, and then examined under a microscope. Larger samples of skin and hair can be placed in a test tube with 10% sodium or potassium hydroxide and warmed in a beaker of water over a Bunsen burner. The test tube sample when dispersed should be centrifuged at 700 *g* for 10 minutes, the supernatant discarded and samples of the deposit examined under the microscope for the presence of mites.

Skin parasites are rarely diagnosed histologically. Sections through lesions caused by burrowing mites may provide information on their developmental stages, but these are usually present in numbers too small to serve for diagnosis. Formol saline (10%) is the usual fixative employed and a haematoxylin and eosin stain differentiates arthropods well.

Where diagnosis is difficult and fungal infection is suspected, special stains, such as Grocott's, will readily stain any hyphae present in the skin and scrapings. The sample should be taken as above and incubated in 10% potassium hydroxide (KOH) incorporating Quink blue black ink, as described by Austwick & Longbottom (1980) (mix together equal parts 40% potassium hydroxide and Parker's blue black Quink). Any spores or hyphae in the preparation are stained deep blue. The best results are obtained if the preparation is allowed to incubate overnight at room temperature in a humid container.

Gut contents

Sholten (1972) described a simple method for staining faecal material for cysts, where the sieved, preserved material was mounted in Meyer's albumin before staining. The method works well with modified trichrome (Alger 1966; Lawless & Burrows 1965; Burrows 1967, 1976) (*see* Appendix B), the sample adhering to the slide and the staining properties remaining unimpaired. The cysts of protozoa require staining to show up the salient features of the morphology to aid identification. In order to show up the organelles and nuclei in the cyst, prepare the sample in 1% aqueous iodine instead of saline. The cytoplasm will stain a pale brown, glycogen masses are darker in colour and the nuclei will be clearly defined. A malachite green stain techique described by Sargeaunt (1962) (*see* Appendix B) is useful for showing up chromatoid bodies, but only in material sedimented by the Ridley formol ether sedimentation method (*see* Appendix B). One drop of the faecal suspension is mixed with one drop of stain on the slide, as with the iodine preparation.

Wet preparations must be made as quickly as possible after the death of the animal. They will ensure that the living trophozoites of many of the protozoa, such as *Giardia*, *Hexamita*, *Trichomonas* and *Entamoeba* sp., are still active and easily recognizable by their distinctive movement; asexual stages of the coccidia can also be distinguished in this way. A loopful of the material mixed with saline and covered with a coverglass should be examined directly under the microscope.

Smears of intestinal contents can be stained with Giemsa after drying a smear of the gut contents suspended in physiological saline with a single drop of serum on a medium-sized coverslip.

Scan the contents of each region of the alimentary tract and surface of the mucosa for the presence of small parasites using a dissecting microscope and a bright light. Any parasites found should be washed carefully and examined microscopically before preparation for mounting. Nematodes are best fixed by dropping into hot (not boiling) 70% ethanol which will both fix and straighten the worms, making identification easier. Some worms such as *Trichuris* must be removed with care if the anterior portion, which is deeply embedded in the mucosa, is not to be damaged. Cestodes must also be encouraged to release their scolices; leaving the tissue and attached worm in tap water until the worm becomes completely flaccid will help. The worms may then be arranged on clean slides for fixing. It is preferable to place the scolex and upper portion of the strobila on one slide and the mid and hind portion of the strobila on another. Place a second clean slide on top of the specimen and press the two together using an elastic band so that the worm is flattened without being damaged. The specimens should then be transferred to a trough containing 10% formol saline and allowed to fix. Specimens may be stored this way indefinitely or stained for mounting.

Transfer cestodes to a solution of acetic acid alum carmine (Schneider's alum carmine) stain diluted 5–10 times in water (*see* Appendix B). The length of time required for overstaining will vary according to the thickness of the worm. Decolourize in acid water or acid alcohol (1% HCl in 70% methanol), clear in cedarwood oil and mount in a suitable mountant such as canada balsam. Small wedges of coverslip or plastic sheet can be used at the edge of the coverslip to raise it, depending on the thickness of the specimen.

Trematodes may be treated in a similar way and mounted with the ventral sucker uppermost in the hermaphroditic species. Schistosomes, however, are not normally flattened. Small flukes and cestodes stain satisfactorily using dilute Delafield's or Erlich's haematoxylin and using acid alcohol to decolourize as before.

Nematodes fixed in hot 70% ethanol may be preserved as they are or in glycerin alcohol (95 ml of 70% ethanol plus 5 ml of glycerol). The alcohol will gradually evaporate and the worm is eventually preserved in pure glycerol. This is a slow process but very effective in clearing small worms without distortion preparatory to mounting. Mount in glycerin jelly and when set the coverslip may be sealed with Xam (BDH Chemicals, Dagenham, Essex, UK). Alternatively, the worms may be fixed in glacial acetic acid for 15 minutes, then transferred to 70% ethanol in which they can be stored. They can be cleared in lactophenol, but care must be taken not to distort the cuticle and passage through the alcohols should be slow. Mount in glycerin jelly as before.

Faeces

(i) Smears

Wet preparations of faecal material can be made for immediate examination for protozoan cysts and worm ova. The sample can be emulsified in tap water or dilute formalin. For large samples the author uses 2.5% potassium dichromate, as this preserves any oocysts present but still drastically reduces bacterial action and thus the strong odour.

(ii) Concentration techniques

First remove the coarser debris from the faecal homogenate by using wet cheesecloth or a sieve of approximately 45 mesh/in^2. The suspension can then be subjected to either flotation or sedimentation techniques.

Flotation methods are usually simple techniques for floating cysts and ova away from the remaining finer faecal debris by use of a solution of high specific gravity such as saturated sodium chloride, sucrose, zinc sulphate, magnesium sulphate or glycerol. Qualitative examination of a stool sample can be rapidly carried out by mixing with a saturated salt solution, centrifuging the suspension, and collecting the surface film on a coverslip (*see* Appendix B). Quantitative methods are best carried out using the McMaster counting chamber, a double glass slide divided into two chambers. These are filled with a sample of a suspension of a known weight of faecal material in a known volume of saturated saline and the number of oocysts or ova rising to the top are counted within the 1 cm^2 area etched into the glass. As the volume examined is 0.15 ml, the number of eggs/weight of faeces can be estimated.

In the sedimentation technique (Ritchie 1948; Ridley & Hawgood 1956; Allen & Ridley 1969) the faecal material is strained and the suspension washed, centrifuged and resuspended in 5% formalin. Ether is added to the suspension and the whole thoroughly mixed before centrifugation at a slow speed for two minutes. The faecal debris is found as a ring at the formalin/ether interface and can be decanted by loosening the debris using a swabstick and emptying the supernatant into a beaker, leaving a small button of deposit at the bottom of the tube ready for microscopic examination (*see* Appendix B). This must be carried out in a fume cupboard to allow safe evaporation of the ether.

If this material is to be stored, the sediment should be suspended in a few ml of 5% formalin in a screw-capped bottle. The author has found this is a far better method than storing even well homogenized samples of whole faeces. The earlier the sedimentation is carried out after collection the better the results. When required for examination, the samples may be centrifuged and examined at leisure. In this state they have been successfully kept for 12 months and more.

Sapero & Lawless (1953) devised a method of preserving faeces and staining any protozoan cysts present. This method is known as the MIF technique (merthiolate-iodine-formaldehyde). Samples from the field can be kept and examined for up to 6 months after collection. The preparation requires a stable merthiolate-formalin (MF) stock solution and 5% Lugol's iodine (*see* Appendix B). It has a definite advantage in the field situation both in the simplicity of its use and the rapidity of fixing and staining cysts and trophozoites in the stool.

Blagg *et al.* (1955) modified this technique by applying the ether sedimentation technique to the MIF preserved sample and named it the MIF concentration technique. The stained cysts and tapeworms were not affected by ether, and comparison with the ether sedimentation, saturated sodium chloride and zinc sulphate flotation methods of concentration appeared to indicate an equivalent, if not greater, retrieval using this combined technique.

Impression smears

Impression smears of tissues or organs can provide a rapid means of finding certain protozoan parasites during their migration. For example, migrating sporozoites of *Eimeria stiedae* can be located in the mesenteric lymph nodes (Owen 1974). The node is carefully halved using a sharp scalpel and the cut surface dried on a tissue before being applied firmly to a clean slide. The smear should be air-dried and fixed in methanol before staining. Whatever the organ, it is essential that it is cleanly cut and that the surface is then dried. Impression smears of the peritoneal cavity of the mouse or other rodents used for serial passage of, for example, *Toxoplasma* or *Encephalitozoon* sp. can be made by applying the centre section of a clean slide to the exposed surface of the viscera.

Blood films and stains

For routine examination, four films should be made as follows:
(i) *For direct examination*: Place a drop of blood, e.g. from the ear, on a clean slide, cover with a coverslip and examine at once. Any living organisms will be seen actively moving among the red cells. At least 30 fields should be examined at a ×40 magnification.
(ii) *Thin smear for staining* with either Leishman's or Giemsa stains (*see* Appendix B): Place a single drop of blood at the end of a clean, grease-free slide. Then place the edge of another clean slide against the leading edge of the drop at an angle of 40–50° and draw swiftly and evenly along the slide producing a thin film, tailing out at the end. Leave to air dry. Fix thin blood films in methanol if they are to be kept for more than a few days otherwise the quality of staining is reduced as time goes on. At least 100 oil fields (×100) should be examined.

(iii) *Thick films for staining* with either Giemsa or Field's stains (*see* Appendix B): Make two thick films by placing a single drop of blood in the centre of each clean grease-free slide and spreading it into a thick oval using the edge of another slide. Thick films do not need fixing; on the contrary the red cells in the film must be haemolysed if the contents are to be seen and identified. Thus it is important that thick films in particular are stained soon after collection. A disadvantage of thick films is that parasites are often less well defined or can be washed off during preparation, but the overriding advantage is that a much larger amount of material can be examined in a very short time. Recognition of parasites in thick films is reasonably easy provided they are properly prepared.

It is preferable that all blood films are stained as soon after collection as possible. If they are to be stored, any immersion oil or dust should be cleaned from the surface with a little xylene. The film should be allowed to dry before being mounted in either euparal or DPX and covered with a cover slip.

Special techniques for microfilariae

Microfiliariae may be identified in thick films prepared as described above. Alternatively the smear can be stained using Mayer's haemalum (*see* Appendix B). Loeffler's methylene blue (*see* Appendix B) can also be used; this is a rapid field method, the stain covering the smear for 15 minutes. In the resultant preparation the sheath is not stained but nuclei are well preserved. Schullhorn van Veen & Blotkamp (1978) developed a quick stain for *Dirofilaria* sp. in which a drop from a Pasteur pipette of 1% aqueous ethanol is spread thinly upon a clean slide and allowed to dry. One drop of sediment from a blood sample haemolysed with saponin can be dropped onto the slide and covered with a coverslip. The microfilariae stretch and die in 2–5 minutes and can be examined any time during the next six hours, after which the larvae become overstained.

Where an infection is suspected but larvae are not demonstrable by direct observation or in thick-stained films, there are methods of concentration of larger volumes of blood. In Knott's test, 10 ml of 2% formalin is added to 1 ml of whole blood (Knott 1939). In this way the red blood cells (RBCs) are haemolysed and the larvae fixed. The sample is either allowed to sediment overnight or centrifuged at 300 rpm for 5 min, the supernatant removed and the sediment examined for larvae.

Alternatively, small volumes of whole blood can be passed through membrane filters (Gordon & Webber 1954; Bell 1967), such as Nucleopore (Dennis & Kean 1971) or Millipore filters (Desowitz & Southgate 1973), and the worms caught in the filter can be stained and examined (*see* Appendix B). There is little or no difference between the results from these two techniques. A comprehensive guide to diagnosis of filariasis has been produced by Denham (1975) and a modified method of counting microfilariae, using a measured amount of blood, is described by Abaru & Denham (1976).

Few blood parasites are found in laboratory rodents because of the lack of suitable vectors. Rickettsiae, such as *Eperythrozoon* and *Haemobartonella* sp., can be found in wild rodents and the latter in larger species such as cats and dogs (Rudnick & Hollingsworth 1959; Maede 1975; Pryor & Bradbury 1975; Simpson *et al.* 1978). In the cat *Haemobartonella* causes fatal infectious feline anaemia; in rodents it is rarely found even in smears, unless the animal is splenectomized when it appears approximately 3–10 days later. So the infection is of little significance unless the animal is severely stressed by age, concurrent infection, such as another blood protozoan, or splenectomy.

Species of *Trypanosoma* commonly found in wild rodents include *T. lewisi* in rats and *T. duttoni* in mice. Transmission is by ingestion of the flea, which is the intermediate host, or its faeces. The parasite can be found in the blood a week later but does not remain longer than 0–14 days. In primates, several species of *Plasmodium* and *Trypanosoma* as well as larvae of filarial worms (microfilariae) may be found in the blood.

SERODIAGNOSIS

Serological techniques for parasite identification are used for a wide variety of parasite-host complexes. Amongst the first tests in general use was the complement fixation test and this and the precipitation tests are still the tests of choice for Chagas' disease and schistosomiasis (Mougeot *et al.* 1978). Many new serological methods have been introduced into the diagnosis of parasitic infection and many reagents are now commercially available, such as antispecies labelled sera for immunofluorescence and enzyme-linked, e.g. horseradish peroxidase, techniques. For the diagnosis of many of the major parasitic diseases of man and animals, complete commercial kits can be obtained although they tend to vary in their efficiency (Kagan 1974).

A few organisms are of particular importance in laboratory animals, such as *Toxoplasma* in all laboratory species and the microsporidian *Encephalitozoon* (*Nosema*) *cuniculi* in rabbits and rodents. *Toxoplasma* is regarded as a highly important organism because of its catholic taste in hosts and as a particular hazard to human health. The most successful early serological test is still in use and is known as the Sabin-Feldman Dye Test (Sabin & Feldman 1948). Suspensions of freshly collected organisms incubated in the sera of infected and non-infected patients differ in their ability to absorb methylene blue dye when stained. The titre of a serum can thus be estimated by a comparative count of the stained and the unstained organisms in the smear. This is a very efficient test but requires the maintenance of a constant supply of both live organisms and suitable human accessory factor.

An indirect haemagglutination test (IHA) was described by Parke (1961) and Lewis & Kessel (1961). Early discrepancies between this and the dye test were mainly due to technical error, but the method introduced by Chordi *et al.* (1964) resulted in 93% agreement. Kits are now commercially available (Wellcome Reagents Ltd, Beckenham, Kent BR3 3BS, UK) and are suitable for use with most laboratory species. A survey of the efficiency of this kit compared with the method of Chordi *et al.* (1964) was carried out on humans and Old World monkeys with satisfactory results (personal observation). An indirect immunofluorescence method (Suzuki *et al.* 1965) is also in current use.

A serological survey carried out by Gannon (1978) using both an immunofluorescence test and peroxidase test revealed that all commercial rabbit breeders in the then Laboratory Animals Centre's Accreditation scheme (Townsend 1969) had a proportion of their breeding animals infected by *Encephalitozoon cuniculi*. This is known to have little pathological significance to its host (Petri 1969; Shadduck & Pakes 1978) but large granulomatous brain and kidney lesions can be seen in histological sections. Sections of the affected organs stained with Giemsa or Erhlich's haematoxylin and eosin demonstrate the presence of cysts but will rarely reveal the presence of spores in lesions. The organisms are not easily identified except by using Gram's stain (Twort & Twort 1923; Ollett 1951) (*see* Appendix B).

A number of serological tests have been described for the detection of *E. cuniculi* and these are reviewed, together with its pathology, by Wilson (1979). Three serological tests have been used in the author's laboratory for screening purposes. The first is an indirect immunofluorescence test, originally developed (Chalupsky *et al.* 1971; Chalupsky 1973) using mouse peritoneal exudate and improved by Cox & Pye (1975) using cell-grown spores. The others are a peroxidase test developed by Gannon (1978) using tissue-culture spores as first described by Waller (1975); and an india ink immunoreaction adapted and developed by Waller (1977) and modified by Kellett & Bywater (1978). The latter test was first developed by Geck (1971) for the diagnosis of enteric pathogens and is now commercially available in kit form (Testman, Box 9020, S-75009, Uppsala, Sweden) for screening for *Encephalitozoon* in rabbits. The main principle of this test is not fully understood, but particles of india ink attach to antibodies in immune rabbit serum. Reaction between antisera and prepared spores is then visible under the light microscope as the ink particles produce a dark ring around each spore.

In this section we have looked at a series of simple screening procedures useful in ascertaining the parasitological status of some of the most commonly used laboratory animals. As a system for post-mortem examination it is a logical progression and can be applied to any species, with minor modification as necessary.

Parasitic infection constitutes a minor part of the abundant soup of microorganisms that the wild animal carries and with which it preserves an uneasy peace. All of the parasitic passengers, however innocuous, produce some reaction in the host and it is in the interest of all workers to ensure that a regular basic screening system is in operation in conjunction with their animal breeding colonies and other animal suppliers.

SECTION 2

ECTOPARASITES (Phylum Arthropoda)

Ectoparasites of laboratory mammals are usually restricted to a narrow range of insects and acarines. The greater variety but smaller numbers found in wild rodents contrasts with the few species but much greater intensity found in the crowded animal room. This intensity, together with the much greater longevity of the laboratory-reared animal, are contributory factors in the syndrome of murine mange.

CLASS INSECTA

ORDER SIPHONAPTERA

A range of flea species has been recorded in the wild-caught rodent (Blackmore & Owen 1968) but few have been observed in the laboratory rodent. In the rabbit only one species, *Spilopsyllus cuniculi*, has been commonly recorded and even that is a rarity, found only in rabbit colonies maintained in outside runs.

This is a very distinctive, totally parasitic order (Richards & Davies 1977). The adult is markedly laterally flattened with a heavily chitinized body and strong, jumping legs. Backwardly directed spines on the body aid rapid movement through the fur, with heavier spines or ctenidia arranged around the back of the head and sometimes around the slender piercing mouthparts. Most have deeply pigmented ocelli and the sexes are easily distinguishable: the upturned posterior region and complex genital apparatus of the male consists of a large, coiled aedeagus and two claspers; the female, on the other hand, has a rounded posterior region with only two small rounded processes, each bearing a short, stout anal sylet. Once cleared in a suitable agent, the spermatheca of the female is easily seen and is sufficiently characteristic to be an important criterion for identification.

The fertilized female lays eggs either in the fur or in the nest of the host, usually the latter. Sometimes as many as 500 are laid in small batches of 20 or so. The ova are oval, pearly white and slightly sticky, measuring approximately 0.5 mm in length. Depending on the environmental temperature and humidity (an optimum of 25 °C and 80% relative humidity) the larvae hatch out in 2–3 days. They are delicate, maggot-like organisms, almost translucent and slightly yellowish in colour with bristles on each of the thirteen segments, the last of which bears two stubby anal struts. There are usually only three larval instars and the mature larva measures approximately 1 cm in length. All stages of this part of the insect's development take place in the humidity and darkness of the nest as they are more susceptible to drying than the adult. The diet consists of debris in the nest and the blood-filled excreta of the adult flea, although there are reports of flea maggots maturing in the fur of the cat and being found filled with fresh blood. This stage lasts for 7–10 days when the larva begins to spin a cocoon, the outside of which is sticky and becomes covered with dust and debris from the host nest, making it difficult to distinguish. This is the most resistant stage in that the pupa is able to survive months, even a year, if left undisturbed. Development of the young adult takes 7–10 days and hatching is stimulated by vibration. The warm proximity of a suitable host attracts the unfed young.

Worldwide, 1400 flea species have been recorded (Smit 1957*a*, *b*), with 53 in the United Kingdom of which 21 have been found on free-living wild rodents. Of these, the most commonly found on the black and brown rat is a large 'combed' flea, *Nosopsyllus fasciatus*, a species which can also be found on the house mouse, field mouse, and field vole. Also common on all these species of rodent is *Ctenophthalamus noblis noblis* and *C. n. vulgaris* of the cat. In more localized areas are other species such as *Nosopsyllus londiniensis* and there has been the occasional report of more exotic species such as the oriental rat flea *Xenopsylla cheopis*.

The most common flea species restricted to the mouse is the tiny *Leptopsylla segnis* (Fig. 2.1, page 18), the 'blind' mouse flea, so called because of its vestigial ocelli. Size varies between the flea species from 1 mm to 3 mm. *Spilopsyllus cuniculi* (Fig. 2.2, page 18) of the rabbit is one of the smaller species, measuring 1.2–1.5 mm in length. Its development conforms basically with other members of the order but differs in that there is a close association between the breeding cycle of the host and its own (Mead-Briggs, 1964*a*, *b*). In the wild rabbit there is a marked seasonal variation, with the highest number of fleas being found between January and April and overall more females than males (Mead-Briggs *et al.* 1975) throughout the year. Both sexes of the flea are influenced by factors in the blood of the prepartum female rabbit and the newborn young during the first five days of life (Rothschild & Ford 1973). The feeding fleas can be found in characteristic clusters in the ears of the host. In all cases there is a strict demarcation of the life cycle into the fur and nest-dwelling stages. The adults only alight on the host to feed, and drop off when engorged. The larval stages remain in the nest where they have been observed feeding on dead nestlings, particularly on those in an advanced state of putrefaction.

The flea is an efficient vector of a large number of disease organisms. Classically the tropical rat flea, *Xenopsylla cheopis*, is the vector of the plague bacillus, *Yersinia pestis*. This is primarily a disease of wild rodents, disease in man following the infection of urban rodents. In tropical and subtropical countries rodent fleas are carriers of endemic typhus.

A common parasite transmitted by fleas in temperate climates is the flagellate *Trypanosoma lewisi* in rats and *T. duttoni* in mice. Fleas are also one of the intermediate hosts of *Hymenolepis* species of tapeworms in the rodent, as they are for *Dipylidium caninum* in the dog. In the wild rabbit *Spilopsyllus cuniculi* is the vector of myxoma virus in the UK, although myxomatosis is rarely encountered in laboratory-bred colonies. Those instances that have been reported were more than likely caused by mosquito transmission.

Few flea species have been recorded in non-human primates (Rahaman 1975; Joseph 1970; Fiennes 1972). *Ctenocephalides felis* from the domestic cat, which have been found in *Macaca radiata* and *Nycticebus* sp., and *C.f. strongylus* in *Colobus caudatus* can all be regarded as passengers. The restless lifestyle of the host would appear to preclude heavy infestation and there is no record in the literature of fleas being vectors of any infectious organism to monkeys.

A somewhat exotic species of flea reported from the gorilla (*Gorilla gorilla*) is *Tunga penetrans* or the jigger flea (Gordon & Lavoirpierre 1962; Matheson 1950). This is one of the smallest of the Siphonaptera and principally a parasite of man. The unfed adult is 1 mm in length, similar in general morphology to other members of the family but less spiny in appearance, belonging to the group of combless fleas. Both male and female are bloodsuckers, though the fertilized female does not leave the host but burrows into a soft portion of the skin, usually the feet. Only the last segments of the abdomen are to be seen protruding from the tissues. Over a period of seven days the ova mature and during the following 8–10 days the eggs are laid and the female dies *in situ*. This is a very painful infestation but, more important, it has a very high risk of secondary bacterial infection.

ORDER PHTHIRAPTERA

The simple life cycle of the louse and its habit of staying close to its host has meant that this parasite has remained a common inhabitant of the fur of many laboratory rodents and, to a lesser degree, of Old World primates.

Suborder Anopleura

The range of species in the suborder Anopleura, or sucking lice, is limited in rodents to two species of the family Hoplopleuridae, *Polyplax spinulosa* and *P. serrata* in rats and mice respectively; and in the rabbit to one species, *Haemodipsus ventricosus*. In Old World monkeys of the cercopithecid group (*Macaca fascicularis*, *M. mulatta*, *Presbytis* sp. and others) a number of *Pedicinus* species have been found representing three genera of the family Pediculidae (Ferris 1916, 1932, 1951; Bedford 1932; Johnson 1958; Benoit 1961).

Members of the suborder Anopleura are morphologically very similar (Farenholtz 1910*a*, *b*, 1916; Ewing 1927, 1933). The female is the larger of the two sexes and usually measures 1.5–2.5 mm in length. The head is long and slender and the mouthparts are adapted for piercing and sucking. There is a marked division into head, thorax and abdomen but the thorax is not segmented. The antennae are three-segmented in the young louse and five-segmented in the adult and the eyes may or may not be present. The three pairs of legs end in clasping organs or claws, usually of a distinctive shape depending on the species. The terminal segment of the female is usually bifurcated, whereas the male is rounded and has a prominent aedeagus. The eggs are laid on the host, usually cemented firmly to the base of the hair. They are generally oval in shape and each has a distinctive operculum with a row of pores round the edge or along the cone. The larvae emerge by ingesting air through the pores, passing it through the body and using the pressure to push-off the operculum lid. The young are similar to the adult except that they are paler in colour and lack the reproductive organs. They then undergo three ecdyses, which takes 1–3 weeks depending on the environmental conditions, before becoming adult; the whole cycle to egg-laying adult takes 2–5 weeks.

In rodents, particularly in laboratory conditions, large numbers of *Polyplax* (Fig. 2.3, page 19) species can be found in aging or sick animals and, when present in large numbers, can cause irritation and possibly anaemia. They are also the most likely vectors for the rickettsial parasites of rodents, *Haemobartonella muris* of the rat and *Eperythrozoon coccoides* of mice. *Haemodipsus ventricosus* of the rabbit is common in the wild but unknown in the better quality laboratory colony. The louse is not known to act as an intermediate host to any other parasite but there has been some discussion as to whether it can act as a mechanical vector of myxoma virus.

Macaca fascicularis, *M. mulatta*, *Presbytis* sp. and many other related species have been found to harbour *Pedicinus obtusus* (Fig. 2.4, page 20). This is a medium-sized louse with long slender claws and a distinctive flask-shaped abdomen. The female measures 2.5 mm in length and characteristically the first three segments of the antennae are fused. The male is 1.75 mm in length with a shorter and stouter head than the female and has five-segmented antennae. The life cycle of these lice is not fully documented but is thought to be typical of the order Anopleura. Many other species of *Pedicinus* have been recorded but none has been associated with the transmission of any infectious agent (Stiles & Hassall 1929; Hopkins 1949; Fiennes 1972).

Pediculus schaeffi is a blood-sucking louse found on chimpanzees *Pan troglodytes* and *Pan paniscus* (pigmy chimpanzee) (Ewing 1932; Kim & Emerson 1968) (Fig. 2.5, page 20). This louse belongs to the family Pediculidae and is almost identical morphologically to the human body louse, *P. humanus* (Nuttall 1919). The female measures 3–4 mm in length, the male 2.25 mm. The life cycle is not documented but it is thought to be similar to *P. humanus*. Unlike the human species, *P. schaeffi* has not been incriminated in the passage of any infectious disease; neither has *Phthirus gorillae* (Fig. 2.6, page 20) from *Gorilla gorilla beringei*, which is morphologically similar to *Phthirus pubis*, the ‘crab’ louse, of man. This is a very distinctive shape, being short and broad with long crab-like legs, the first pair slightly smaller than the others. The head is longer than broad with large eyes on lateral protruberances and not constricted into a neck. The thorax is broad with its segments fused and the short abdomen also has the first three segments fused, leaving four remaining segments, giving an overall appearance not unlike a small crab. The female has a total body length of 2.2–2.25 mm but details of the male are not known. The life cycle is thought to be similar to *P. pubis* of man, a species which lives in the coarse body hair, survives only a few weeks and lays approximately 25 eggs during that time. Like all species of louse this *Phthirus* is host-specific, tends to be less active than other species and survives only a short time away from the host.

Suborder Amblycera

The most common examples of the suborder Amblycera (previously Mallophaga), or chewing lice, can be found on the guinea pig, which is host to three species. The most frequently found is *Gliricola porcelli*, or the slender guinea pig louse (Fig. 2.7, page 21). The female measures 1.68×0.27 mm and the male slightly less. Typically the head is longer than broad but still heavier than the head of blood-sucking lice in order to accommodate the cutting mouthparts, which, in this species, are armed with minute serrations and teeth and form part of the hypopharyngeal chitinization. The four-segmented antennae slot into deep antennal fossae. As in the Anopleura, the eggs are cemented to the base of the hair and, in the guinea pig, particularly to the fine hair around the back legs and anus. The eggs are oval and white with a very distinctive operculum, which appears to be perched like a hat on the top of the egg and the whole is cemented very untidily to the hair.

Another common louse of the guinea pig is *Gyropus ovalis* (Fig. 2.8, page 22). This is slightly shorter but much broader than *Gliricola*, the female measuring 1.03×0.52 mm; the male is slightly smaller. The head is wide with projecting posterior margins, giving it a very heavy appearance. Each of the first pair of legs is clawed and the second and third pairs curve inwards for clasping.

Less common now is the largest of the guinea-pig lice, *Trimenopon jenningsi* (Fig. 2.9, page 23). The female measures 1.72 × 0.68 mm and the male 1.6 × 0.64 mm. They are dark brown in colour, the head broad with dorsal expansions and the four-segmented antennae lie in grooves partially covered by the dorsal expansions. The mesothorax is small and fused with the prothorax and the abdomen is broad and oval in shape. The legs are unspecialized and each bears a two-clawed tarsus. The eggs are large and can be seen with the naked eye. They are goblet-shaped and strongly patterned, giving a basket weave effect, and can be found either singly or in groups scattered throughout the fur. These are unlike the eggs of *Gyropus ovalis*, which are regular in shape with a neat circlet of air holes around the edge of the operculum and which are generally laid on the stronger hairs of the neck.

As with the Anopleura, the young lice emerge as pale replicas of the adult, undergo three ecdyses and development from egg to adult takes place within two to three weeks. These parasites have not been incriminated as vectors of any infectious organism. They feed entirely on epidermal scales although sebaceous secretions form a part of the diet of *Gliricola porcelli*. In a heavy infestation of the latter, the hair may become brittle but there is no evidence that the louse actually feeds on the hair. *T. jenningsi* has been found with some host blood in the intestine, but this is thought to have been accidental through feeding in the area around sores or wounds.

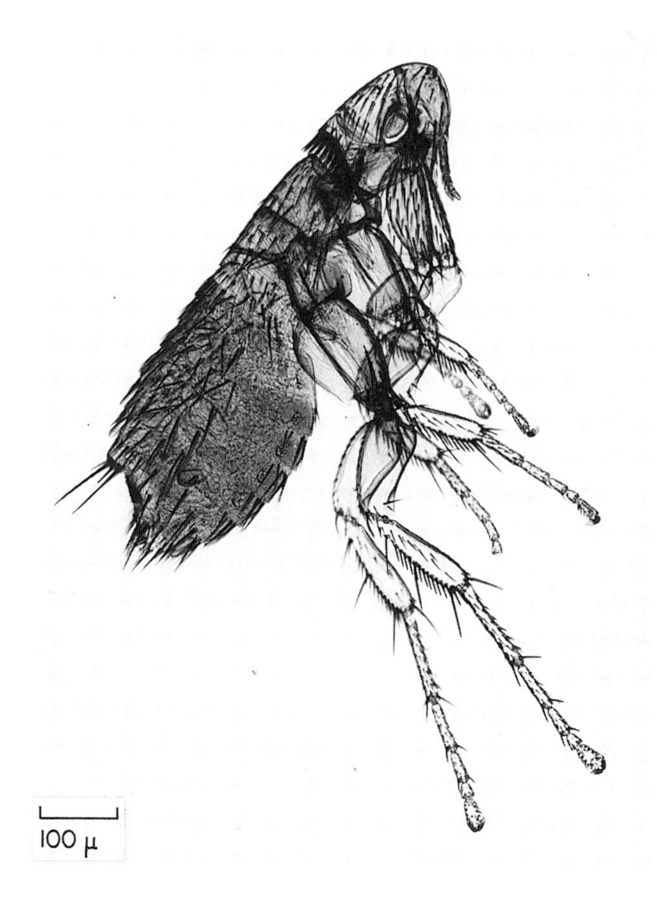

Fig. 2.1 *Leptopsylla segnis*, the 'blind flea', from a wild mouse.

Fig. 2.2 *Spilopsyllus cuniculi*; adult female flea from a rabbit.

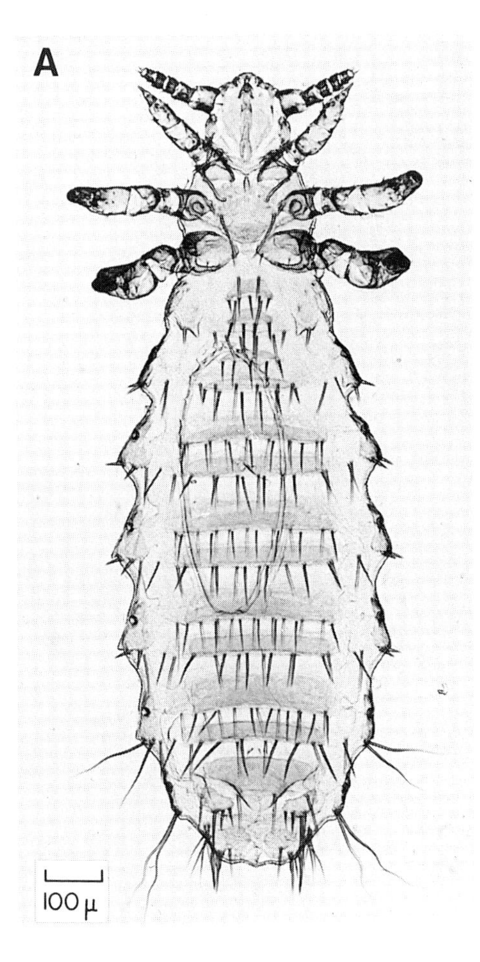

Fig. 2.3 *Polyplax spinulosa*, a sucking louse from a rat. **A**: Female. **B**: Male.

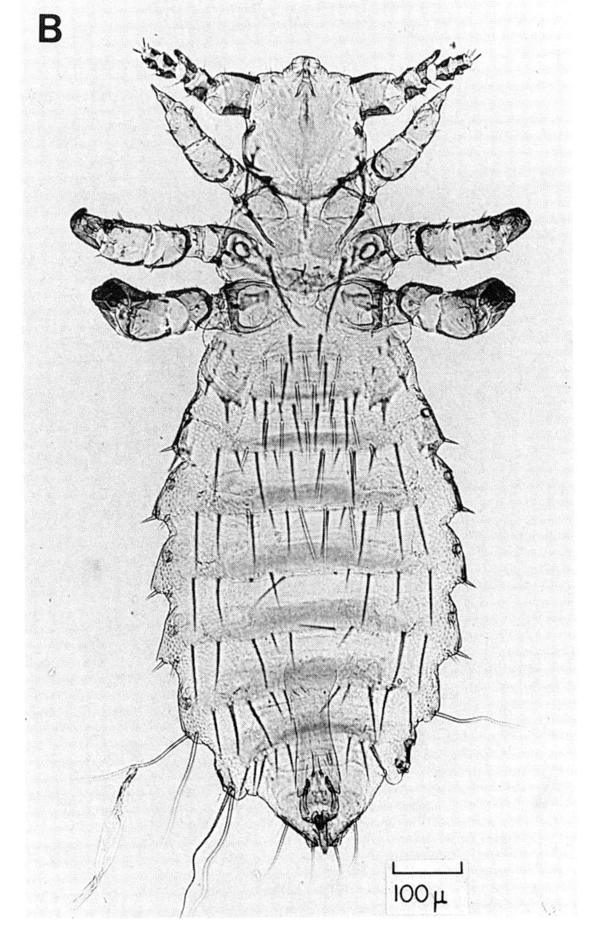

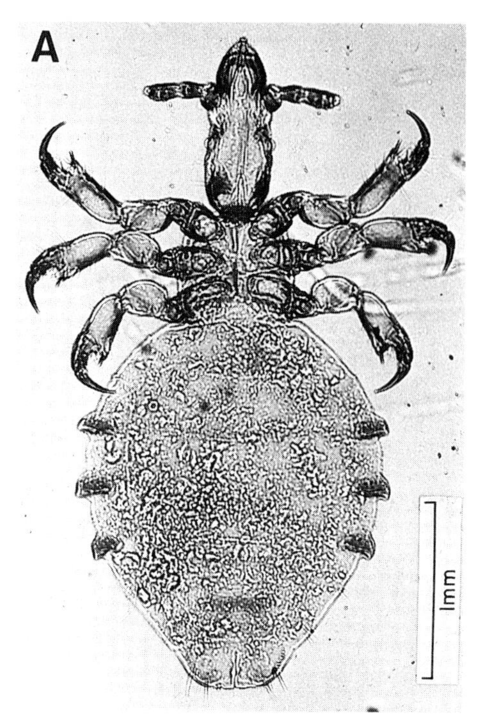

Fig. 2.4 *Pedicinus obtusus*, sucking louse from a rhesus monkey. **A**: Adult. **B**: Group of individuals on the hair of the host.

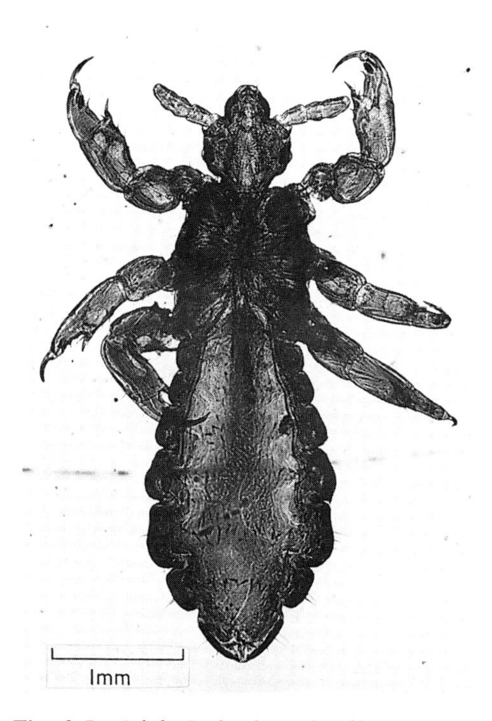

Fig. 2.5 Adult *Pediculus schaeffi*, a blood-sucking louse, from a chimpanzee.

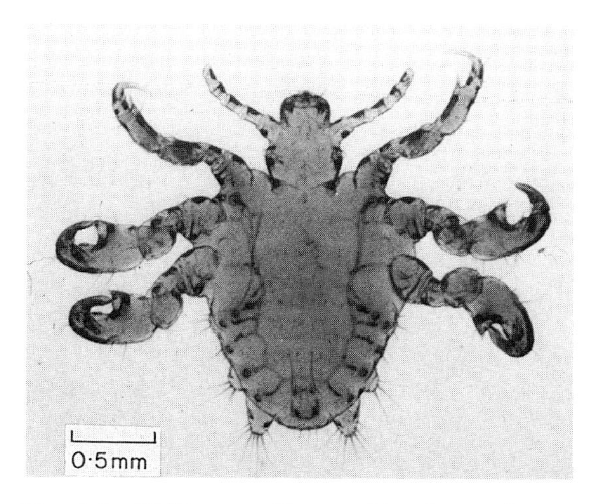

Fig. 2.6 *Phthirus gorillae*, sucking louse, from a gorilla.

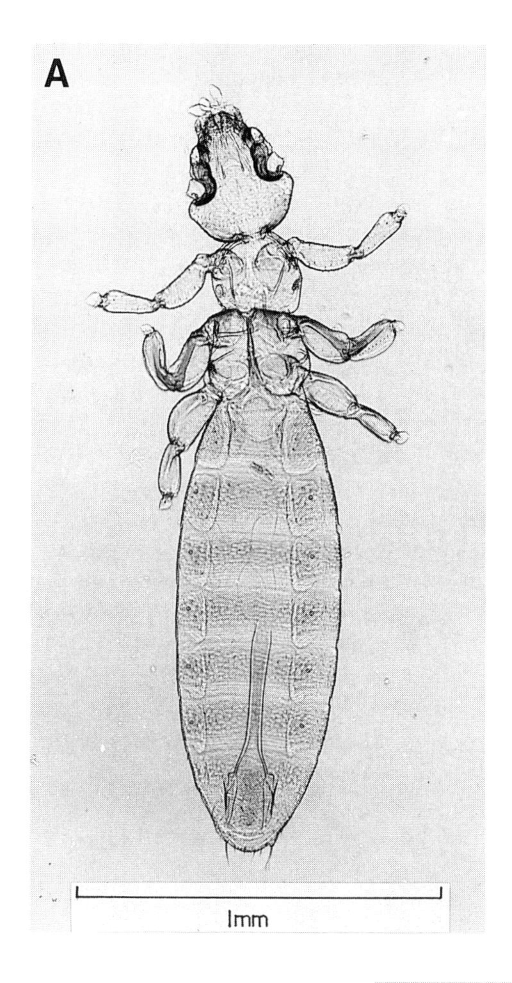

Fig. 2.7 *Gliricola porcelli*, the 'slender guinea pig louse'. **A**: Adult female. **B**: Egg with distinctive operculum.

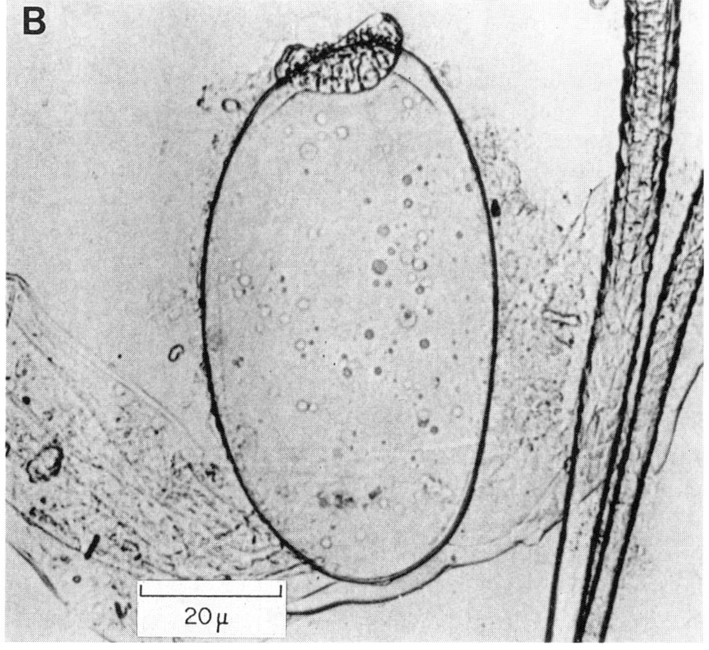

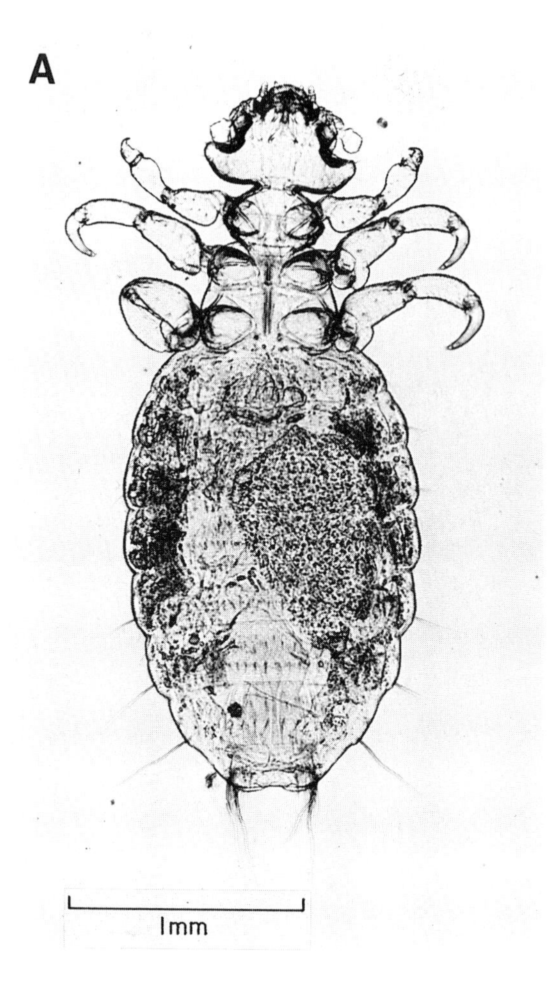

Fig. 2.8 *Gyropus ovalis*, chewing louse from the guinea pig. **A**: Adult female. **B**: Characteristic egg cemented to the hair of the host.

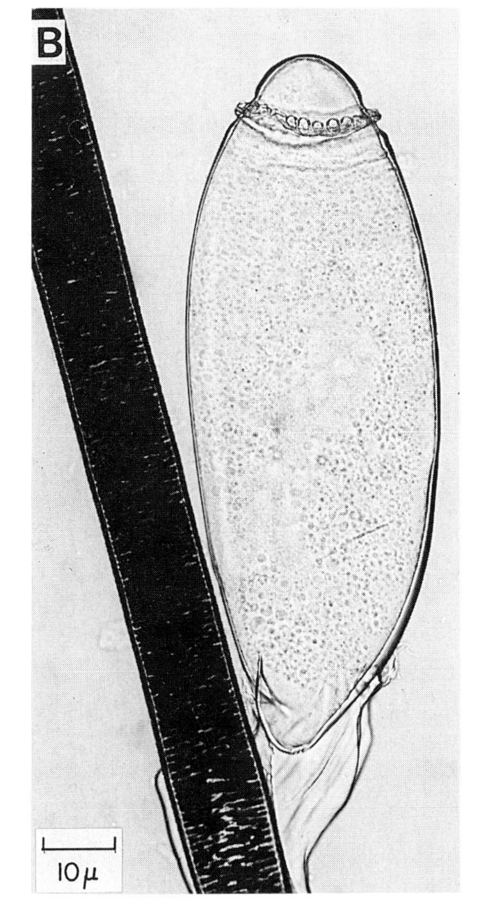

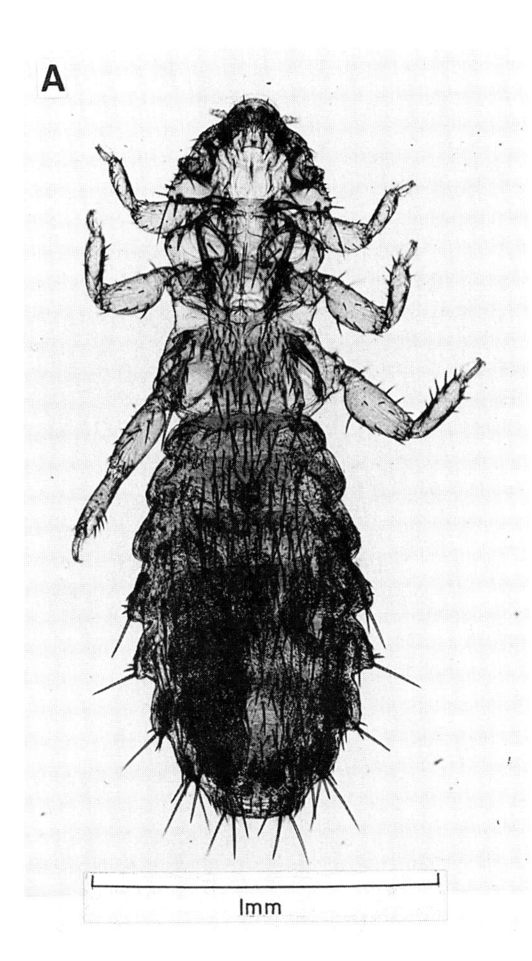

Fig. 2.9 *Trimenopon jenningsi*, the largest guinea pig louse. **A**: Adult female. **B**: Distinctive patterned egg.

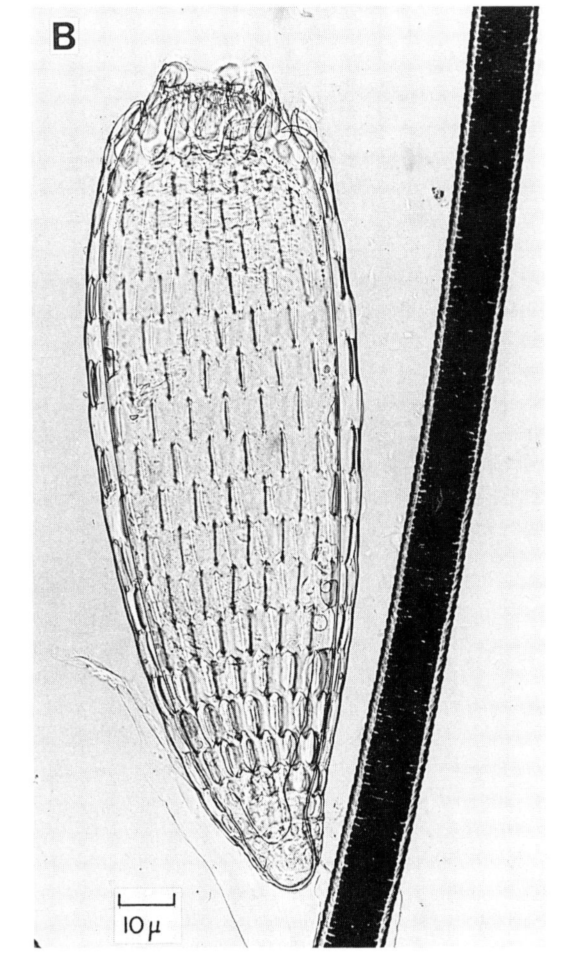

CLASS ARACHNIDA

ORDER ACARINA

Mites are the most common cause of mange in laboratory animals, which can be a considerable problem in the special circumstances of the animal house. The prolonged lifespan and dense population of rodents result in a massive accumulation of the parasite and serious disease problems in long-term experimental animals.

The Arachnida differ widely in appearance from the insects considered previously (Baker *et al.* 1958) in that there is no differentiation into head, thorax and abdomen. Either there is a division between the cephalothorax and abdomen, as in the case of some spiders, or, as in all ticks and some mites, the body is completely fused. There are four pairs of ambulatory limbs and six pairs of appendages at the apex of the cephalothorax. Two pairs are concerned with feeding: the chelicerae, which are either pincer-shaped or modified for piercing; and the palps or pedipalps which are leg-like in appearance and, with the epistome and hypostome, make up the principal structure of the mouthparts. There are no wings and the abdomen has no appendages, only the terminal opening of the anus and a ventral genital opening. There is a wide variety of forms, depending upon their host's mode of life, amongst those species in close association with mammals. They are mostly oviparous and the larva resembles the adult, except that it has three pairs of legs and no reproductive organs. The larval stage is followed by three nymphal stages before sexual maturity is reached.

Fur-dwelling mites

The most common mange-producing mites in laboratory animals are the fur-dwelling species. In rats and mice this is either *Myobia musculi* (Fig. 2.10, page 31) or the very closely related species of *Radfordia* (Fig. 2.11, page 31). These are pearly white mites seen in close association with the base of the hair. The female measures 390–495 μ × 190–260 μ and the male is slightly smaller. The first pair of legs is closely applied to the slender chelicerae and modified to clasp firmly to the hair during feeding. The remaining three pairs of legs are simple five-jointed structures with a terminal claw. There is a markedly characteristic body shape, with the lateral margin extending out into lobes between the limbs. Differentiation between *Myobia* and *Radfordia* is easily made by noting the number of claws on the second pair of legs: in *Myobia* there is one and in *Radfordia* there are two (Ewing 1938). The eggs, as in all fur-dwelling species, are firmly cemented to the base of the hair. Under good conditions the period of time from egg, through larva, nymph to adult can be as little as 12–14 days. They feed by piercing the superficial epidermal layers of the skin and although this produces a thickening of the epidermal layer (Watson 1961), clinical signs may not be apparent provided there are no concurrent infection or stress factors.

Equally common in the mouse is the mite *Myocoptes musculinus* (Fig. 2.12, page 32), also found in the Djungarian hamster and the Russian steppe lemming. Similar in general size to *Myobia*, the female is pearly white in colour and measures 330–380 μ × 155–200 μ with the third and fourth pairs of legs heavily chitinized and adapted for clasping. The male is smaller and darker in colour with only the third pair of legs developed for clasping, the fourth pair being enlarged, backwardly directed and terminating in stout claws.

In both these species it is debatable whether the mites are the sole cause of the mange syndrome. A healthy host will tolerate a heavy infestation without apparent ill effects; the increase in numbers is usually associated with age or other stress factors such as pregnancy or vitamin A deficiency. Irritation results in hair loss and an obvious reddening and thickening of the skin. If sores are produced then a secondary bacterial infection may follow (Fig. 2.13, page 32).

In rabbits, mange has been known to be produced by a mite, *Cheyletiella parasitivorax* (Fig. 2.14, page 33), which is very commonly found in large numbers in the fur. This is a medium-sized mite, the female measuring 450 × 200 μ and the smaller male 320 × 160 μ. They are very active, with a yellowish-white body, making detection fairly difficult on albino animals. The short, sturdy limbs lack a tarsal claw but the narrow empodium carries a distinctive comb-like row of hairs. Perhaps the most distinctive character is the palpi which are short and broad with a very prominent claw curving inwards, giving a pincer-like appearance.

Other members of the family Cheyletidae are free-living predators and *C. parasitivorax* differs little in its general morphology. The life cycle, although not well studied, is thought to resemble that of the free-living members. It was once thought that the same species was found on cats and dogs, but these are now named *C. blakei* and *C. yasguri* respectively (reviewed by Bronswijk & de Kreek 1976). Man is a very rare transient host for this parasite, most cases having been reported from humans in contact with affected cats or dogs. Under the stress of experiment, mange may appear in rabbits, usually on the shoulders and back and consisting of dry, flaky skin with extensive hair loss. A small skin scraping will reveal the presence of a number of mites.

Another common mite of rabbits is *Listrophorus gibbus* (Fig. 2.15, page 33). This species tends to remain on the hair shaft and is not thought to have any pathological significance whatever. It is, however, often present in considerable numbers and for that reason cannot be ignored (Gunther 1942). It is a small mite, slightly laterally compressed, closely attached to the shaft of the hair by means of the palpal coxae which form two chitinous flaps. The mite feeds on the secretions of the sebaceous glands and scales from the surface of the hair. The heavy chitinization of some parts of the body give these mites a dark brown appearance and the eggs, laid singly, are cemented to the hair usually further up the shaft than those of *C. parasitivorax*, which are usually found closer to the skin surface.

Closely related to *Listrophorus* is a fur-dwelling mite of the guinea pig, *Chirodiscoides caviae* (Fig. 2.16, page 33), a small mite, usually twice as long as broad. The female measures 460–500 μ, the male slightly smaller. The first two pairs of legs are heavily chitinized, long and adapted for wrapping around the hair shaft, whereas the third and fourth pairs of legs are unspecialized. In the male the last pair of legs is considerably longer than the third and each bears a long blunt hair laterally. The long slender egg is laid half way up the hair shaft and the life cycle from egg to adult is similar to other mite species. The young stages, like the adults, seem to spend the majority of their time away from the skin of the host. Again this mite is not thought to have any particular pathological significance.

Surface-dwelling and burrowing mites

The mites that most consistently produce mange lesions are those that spend their life in closest contact with the skin of the host. The majority of these mites belong to the families Psoroptidae or Sarcoptidae. One species of the Psoroptidae, *Psoroptes cuniculi*

(Fig. 2.17, page 34), is more commonly known as the ear canker mite of rabbits. The female is a large, active mite measuring 513–1020 μ in length by 502–610 μ in width, and the male approximately 620 × 405 μ. They are surface-dwellers and can be seen by the naked eye as oval bodies moving quite rapidly inside the ear pinna. Early infestations can be diagnosed using an auroscope when the mites can be seen deep inside the auditory meatus. They are fairly well chitinized, the robust dark legs ending in a characteristic three-jointed pedicel with a bell-shaped terminal sucker. The ovigerous female has a conspicuous U-shaped genital opening and the third and fourth pairs of legs have no suckers but both bear setae, those on the third pair of legs being particularly long. Suckers are on the first three pairs of legs in the male, not on the fourth, which is also considerably shorter (Fain 1968; Sweatman 1958). The characteristic body shape and limbs together with the three-jointed pedicel are among the basic criteria for differentiating the Psoroptidae from the Sarcoptidae.

The eggs are laid on the skin and the period of development from larva through nymphal stages to adult can be completed in 8–9 days. Initially there is little apparent effect on the host, but as the numbers increase so does the irritation caused by the mites as they pierce the skin surface. This results in a host response which is manifested by a serous exudate which gradually fills the entire external auditory meatus. Reaction may spread to the outer edges of the ear or even to the face in cases where the condition has been neglected. All the exudate must be removed if treatment is to be successful and, at this stage, such treatment would require a general anaesthetic.

Morphologically identical species of *Sarcoptes scabiei* (Fig. 2.18, page 34) are found on a wide range of rodents, domestic animals and man (Baker *et al.* 1967). These almost spherical, pearly white mites are usually found in pockets or burrows in the epidermis of the host. The female measures approximately 300–400 μ in length and the male up to 250 μ. The finely striated cuticle is grooved with small wedge-shaped scales and the limbs are short and thick-set. The first two pairs of legs in both female and male and the fourth pair of legs in the male bear suckers on unjointed pedicels. The other legs in both sexes have long distinctive bristles. The eggs are laid within the tunnel in the epidermis. During a period of three to four weeks a total of 30–40 eggs are laid (Fain 1968). The larvae may remain in the burrow or may burrow their way on to the surface. Each of the two nymphal stages may also construct their own burrow but these nymphs are rarely found on the surface. Fertilization of the female takes place in the burrow and development from egg to adult takes 10–14 days. The burrows are difficult to detect and even a few mites may cause considerable irritation to the host.

A detailed study of the life cycle of *Sarcoptes* and a closely related species in rats, *Notoedres*, was first carried out by Gordon *et al.* (1943), who found that there is a division of populations in both infestations: a surface-dwelling population of larvae, nymphs and adult males; and a sub-surface population of ovigerous females and eggs and those other stages that have not yet left the parental tunnel. Thus the spread of infection is principally through the transfer of larvae and nymphs. In simian primates the infection usually first appears on the neck and shoulders but can be found extending to the trunk and extremities (Delorme 1926; Brug & Haga 1930; Leerhoy & Jensen 1967; Lavoirpierre 1970).

Notoedres muris, found in the wild rat (Fig. 2.19, page 35), prefers the epidermal tissue of the ears and face and, in very heavy infestation, can spread to the legs and genitalia as a result of grooming. Both sexes are globose, white in colour with reduced legs, the first two pairs of which bear suckers on long unjointed pedicels. The male also bears suckers on the fourth pair of legs, and the remaining limbs have fine setae. The female is a little smaller than *Sarcoptes*, measuring 300 × 200 μ and the male 200 × 160 μ. As described previously, the ovigerous female remains in the burrow, proceeding to enlarge it as she lays eggs. The time of development is approximately 19–21 days. Lesions first appear around the rim of the ear where the skin becomes thickened and reddened; a crust appears followed by the characteristic warty papules. In severe cases the whole of the pinna may become eroded. Morphologically similar species have been reported in the rabbit (Adams *et al.* 1967), in the cat (Davies 1951; English 1960) and in the golden hamster (Enigk & Grittner 1951; Bàies *et al.* 1968).

Other genera of the family Sarcoptidae have been found on a variety of primates. *Prosarcoptes pitheci*, for example, has been found in *Papio papio* and *Cercopithecus* species and differs from the genus *Sarcoptes* by the presence of stalked caruncles on all four pairs of legs in both sexes, the dorsal spines evenly distributed and similar, and a slight variation in size (Fain 1968). Other species consist of *P. faini* in the Chacma baboon and *P. scanloni* found in *Macaca fascicularis* and very similar *Kutzercoptes grunbergi* in *Cebus capucinus* (Philippe 1948; Lavoirpierre 1960; Zumpt 1961; Smiley 1965; Kutzer & Grünberg 1967).

Trixacarus caviae (Fig. 2.20, page 36) is a sarcoptid mite found in the guinea pig (Fain *et al.* 1972), occasionally causing mange in breeding colonies (Beresford-Jones *et al.* 1976). The female measures $189 \times 144\,\mu$ and the male $135 \times 102\,\mu$. Both are typically white in colour, globose with fine striations in the anterior half of the body with numerous elongate scales. The limbs are short and in both sexes the first two pairs of legs bear suckers on unjointed pedicels; in the female the third and fourth pairs of legs bear long setae, whilst in the male only the third pair have setae. Details of the larval and nymphal stages are available but not of the timing of the life cycle. Many of the old standard textbooks include descriptions of a form of sarcoptic mange attributed to *Sarcoptes scabiei*, but this was most probably due to *Trixacarus caviae* (Lavoirpierre 1960). The clinical signs are very similar: irritation and scratching is followed by thickened and wrinkled skin and hair loss with lesions that subsequently become secondarily infected if the condition is not treated. The most commonly affected areas are the neck and shoulder region, the lower abdomen and thighs. They respond well to treatment but in most cases reported there would appear to be factors causing stress which exacerbate the situation.

Follicle-dwelling mites

Other mites to be found commonly in the skin, which are not burrowing but live in the hair follicles and sebaceous glands of a wide variety of mammals, are *Demodex* sp. and *Psorergates* sp. In the laboratory rodent *Demodex* is not usually considered to be of any pathological importance, but in the Golden hamster (*Mesocricetus auratus*) it has been associated with a severe mange under certain circumstances (Nutting 1961, 1965; Estes *et al.* 1971; Owen & Young 1973).

D. criceti is a very small mite, weakly chitinized, with the almost vestigial legs typical of the species. It is short, the female measuring $103\,\mu$ in length and the male $87\,\mu$. The stumpy limbs are placed evenly along the podosoma. The mites are usually found in epidermal pits in the skin where they lay their thin-walled eggs. These measure $38 \times 20\,\mu$ and when the larva emerges its three pairs of legs are little more than plates with two tubercles. Nymphal stages do follow, though it is not known how many or the minimum length of the development period, but multiplication is possibly very slow.

The other *Demodex* species found on the hamster, *D. aurati* (Fig. 2.21, page 36), is found in the hair follicle and differs from *D. criceti* morphologically by its much more elongated shape, the female measuring 191 μ in length and the male 183 μ. The spindle-shaped eggs measure 65 × 17 μ and are laid in the hair follicle. The diet of *D. aurati* would appear to be the epithelial cells of the follicle and sebaceous gland; whereas the presence of *D. criceti* on the skin surface would indicate that the surface epithelium consitutes the diet of this species. *D. aurati* can be found in fairly large numbers without any apparent ill effects, but where mange is patent a scraping will reveal an enormous mass of mites in the affected area. Clinically, the skin is dry and scaly with many scabby lesions accompanied by patchy hair loss over the rump and back.

Demodicosis has also been reported in the mongolian gerbil (*Meriones unguiculatus*) (Schwarzbrott *et al.* 1974). This was reported from an old male animal and the description of the mange is essentially similar to that found in the hamster; the name *D. merioni* was given by the authors. Species have been described in rats, *D. ratti* in the brown rat (*Rattus norvegicus*) (Hirst 1917, 1919, 1922), *D. nanus* in the black rat (*Rattus rattus*) (Hirst 1915, 1919) and possible demodicosis with *D. nanus* in laboratory rats (Wallberg *et al.* 1981). In the dog, where *D. canis* may be present in large numbers, it is associated with severe mange. Two species are found in man, *D. folliculorum* and *D. brevis*, but only rarely does their presence result in demodicosis. The same applies to those species described from non-human primates such as *Rhinodex baeri* and *Stomatodex galagoensis* from the bush baby (*Galago senegalensis moholi*) described by Fain (1959*b*), and *D. saimiri* from the squirrel monkey (*Saimiri sciureus*) (Lebel & Nutting 1973). *Demodex* species have been noted from a range of other non-human primates, both Old and New World, but they have not been described fully.

The other follicle-dwelling mite that is less common in laboratory animals in the United Kingdom is *Psorergates simplex* of the mouse. It has been reported in continental Europe and the United States, with high incidences reported in the former (Flynn 1955) but only a few reports from the UK (Beresford-Jones 1965; Cook 1956). This is a very small mite, spherical in shape, the female measuring 189 × 162 μ with two pairs of long setae at the posterior end, and the male 167 × 116 μ with only one pair of setae. The legs are arranged radially and at the base of each short stout femur is a very distinctive medially directed spine. Each limb bears a pair of simple tarsal claws and a pad-like empodium. The detailed life cycle is not completely known; however, all stages of the parasite, egg, nymphal stages and adults can be found in the nodular cysts which are readily observed at post-mortem examination when the skin is reflected. They are seen as round, white bodies approximately 2 mm in diameter and consisting of a flask-shaped invagination of the epidermis filled with sebaceous material. A detailed description of the stages has been made in the case of the very closely related species, *P. ovis*, of sheep (Davis 1954). Mites of this genus have been found on a range of African primates: *Cercopithecus aethiops* and *C. pygerythrus* (Zumpt & Till 1954, 1955); *C. mona (mona)* (Lavoirpierre & Crewe 1955); the sooty mangabey, *Cercocebus (torquatus) atys* (Sheldon 1966); and the patas monkey, *Erythrocebus patas* (Raulston 1972).

In mice, infection with *P. simplex* appears to have no clinical effect unless the parasite is transferred to what is apparently an abnormal site, as in the case reported by Cook (1956) where the mites had been transferred to the edge of the ear, possibly by scratching, and where a thick yellow crust appeared on both sides of the rim as a result of irritation. In the patas monkey the appearance of infection was a series of periorbital yellow patches which gradually coalesced, and large numbers of mites were found in scrapings taken from the site. In both mice and monkeys the lesions regressed and disappeared after treatment.

Mites of the lung and nasal cavities

Parasitic mites of the lung and nasal cavities are common among Old World monkeys (Innes 1969: Hull 1970; Innes & Hull 1972; McConnell *et al.* 1972), particularly macaques, and incidences of 100% have been recorded from some surveys carried out on *M. mulatta*. *Pneumonyssus simicola* (Fig. 2.22, page 37) is found in a wide range of other species such as *M. nemestrina*, *Mandrillus sphynx* and *Theropithecus gelada*. All stages of the mite are to be found in nodules in the lung. The adults are ovoid, measuring 600–1600 μ in length and 300–500 μ in width. They are pale in colour, whitish or cream with darker chitinized limbs. The sluggishly moving parasites can be seen easily with the naked eye. The basic life cycle is thought to be similar to that of other Mesostigmata but some stages may be lacking and in most groups the larvae are born viviparously.

It is very difficult to diagnose infection by general examination. The movement of the mite in the lungs can cause irritation and coughing, but there is no way of assessing the amount of damage caused to the lung in a live animal. All wild rhesus monkeys are affected to some degree on capture and appear to tolerate considerable numbers of the mites without any ill effects. Young born in captivity have been successfully kept clear of the mites by dusting of the mother's pelage with a residual acaricide. The nodules in the lung vary considerably in size from pinhead to that of a small pea. In a section through the lung the parasitic nodules can be seen with areas of inflammation surrounding the nest, and distinctive black pigment can be seen around the lesion and in areas of apparently unaffected lung (Knezĕvich & McNulty 1970).

Mites affecting the nasal cavities, *Rhinophaga elongata*, have been described in some primates (Fain 1958, 1959*a*; Coffee & McConnell 1971). These are usually found on the surface of the mucosa but they can be buried in nodular lesions which sometimes penetrate the subadjacent bone. The most distinctive feature of this mite is elongation of the body. The first three pairs of legs are grouped in the anterior part of the body while the fourth pair is situated half way along the slender, thread-like body. The length of the idiosoma varies between 4740 μ and 5203 μ, the width between legs II and III being 569–575 μ. The dorsal shield is long and oval, measuring 518–575 × 268–288 μ, is punctate and bears a median sclerotized plate. Two pairs of setae are situated on the median plate and the remaining four laterally. Ventrally, there is a punctate sternal shield with three pairs of setae which lies over a sclerotized plate with four lateral extensions. The gnathostome is 211 μ long and 94 μ broad, the hypostome has two pairs of setae and the deutosomal groove seven pairs of teeth. The legs are roughly equal in length but with claws II and III slightly longer and stronger than I and IV. In the original discovery the female was found with its anterior third embedded in the nodular lesions.

Like *Pneumonyssus*, the life cycle of *Rhinophaga elongata* is not fully documented but it is thought that it cannot live for long away from the host and transmission must therefore be direct. Mites embedded in nodules may have serious pathological effects on the host.

Infestation of the nasal cavity by mites is not usually a problem of laboratory rodents. However, infestation of hamsters by the field vole mite, *Spelorodens clethrionomys* (Fain & Lukoschus 1968), was reported by Bornstein & Iwarsson (1980) (Fig. 2.23, page 37). This was a commercial colony of Syrian hamsters intended to supply animals for inhalation experiments, but when examined 90% of adults and young were found to be infested. The mites are milky white in colour and oval in shape. The female measures 300–360 μ by 114–198 μ but the male has not been described. The cuticle is finely striated and there is a dorsal shield with thicker sclerified lines. The four pairs of legs have very characteristic plumose setae, oval in shape and particularly large on the first pair of legs. All tarsi have setae and all carry distinct apical claws. This appeared to be an isolated outbreak, these and other animals all coming from the same supplier and none showing any apparent clinical signs of disease.

ORDER PENTASTOMIDA

The pentastomids are a highly aberrant order of arthropods that live in the respiratory tracts of many carnivorous animals, usually snakes, and the intermediate host for larvae is usually a mammal.

Porocephalus clavatus, found in South America, is a parasite of *Constrictor constrictor* and the larval stages are found in primates such as *Saguinus fuscicollis* and *S. nigricollis*. The adults are worm-like in shape, pseudosegmented by characteristic chitinous rings. The mature adult measures 3–12 cm in length with four hooks, two on either side of the mouth. There are no other appendages and the sexes are separate. The female uterus becomes gravid with ova which are laid in the lungs of the primary host and are eventually swallowed and passed out in the faeces. The egg is ingested by the mammalian host as a contaminant of the drinking water. On reaching the gut it hatches and the four-legged larva emerges. It burrows through the intestinal wall and makes its way to the abdominal cavity where maturation to the nymphal stage takes place in one of the organs, usually lung or liver, and results in an organism similar to the adult except for size and lack of sexual organs. Macroscopically the larvae can be seen coiled below the liver capsule. Further development only occurs if the intermediate host is eaten by a suitable definitive host. These nymphs are capable of invading a wide variety of hosts and in man the invasion results in a clinical condition known as porocephalisis.

P. crotali is a parasite of the rattlesnake and the nymphs have been described in rodents such as the deer mouse and the cotton rat (Layne 1967; Self & McMurry 1948). *P. subulifer* (Fig. 2.24, page 38) is found in the file snake, *Mehelya*, and the nymphs in African primates, *Cercopithecus callitrichus* and *Galago senegalensis* (Stiles & Hassall 1929; Graham 1960; Fain 1961, 1966; Myers & Kuntz 1965; Reardon & Rininger 1968; Self & Cosgrove 1968, 1972).

Armillifer armillatus (Fig. 2.25, page 38) is similar in most respects to *P. clavatus*. It is, however, confined to tropical regions, particularly Africa, where it is common in a wide range of monkey species (Bezubik & Furmaga 1960: Fain 1960, 1966; Henschele 1961; Fiennes 1967).

Another related pentastomid, *Linguatula serrata*, is usually a parasite of canines and only rarely man. It has, however, been reported from a wide range of mammal hosts including the South American monkeys *Callicebus caligatus* and *Saguinus nigricollis* (Stiles & Hassall 1929; Schacher *et al.* 1965; Nelson *et al.* 1966; Cosgrove *et al.* 1970). Morphologically the organism is elongated and leaflike in shape, with a transversely striated cuticle and two pairs of retractile hooks at the anterior (broad) end. The female measures 8–13 cm, and the male 1–2 cm. The eggs measure 90×70 μ. Eggs discharged from the respiratory tract of the primary host reach the secondary host via contaminated food or water. On being ingested the larva burrows its way to the lymph glands where it develops to the nymphal stage. This is whitish in colour and measures 5–6 mm long, very similar to the adult but sexually immature. Infection of the definitive host is by ingestion of the mature cyst containing the nymph. The adults live approximately 15 months. This is a rare parasite of both monkeys and man. In the latter it is reported to cause a disease syndrome known as *halzoun* (Schacher *et al.* 1965).

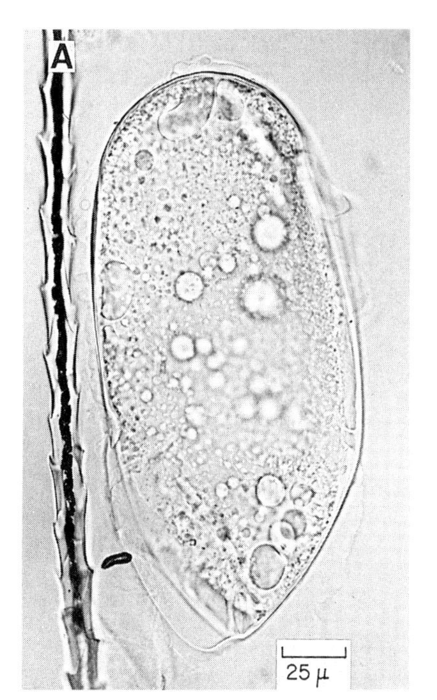

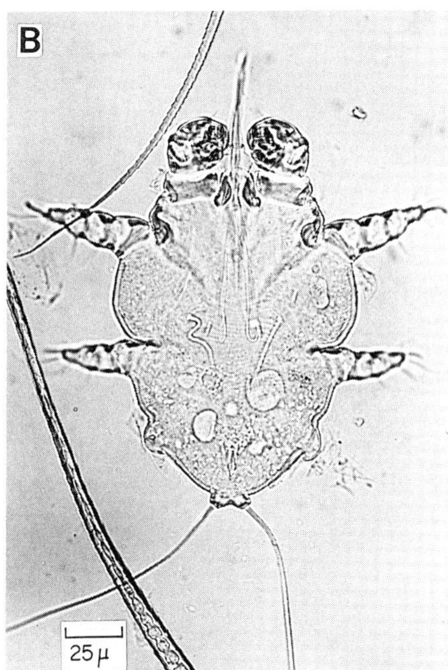

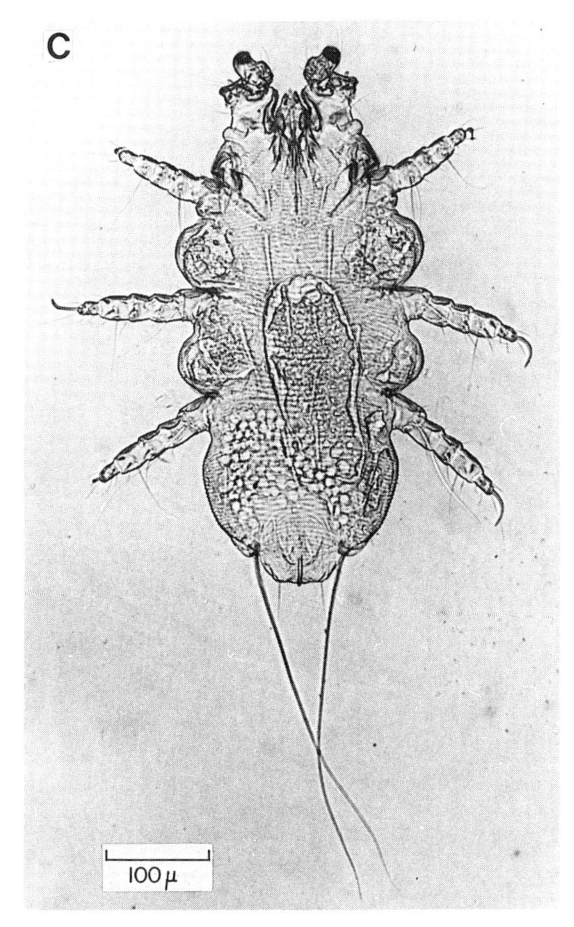

Fig. 2.10 *Myobia musculi*, mange mite from a mouse. **A**: Egg cemented to the hair. **B**: Typical six-legged larva with the first pair of legs modified for gripping the hair and showing stylet like mouthparts. **C**: Adult female.

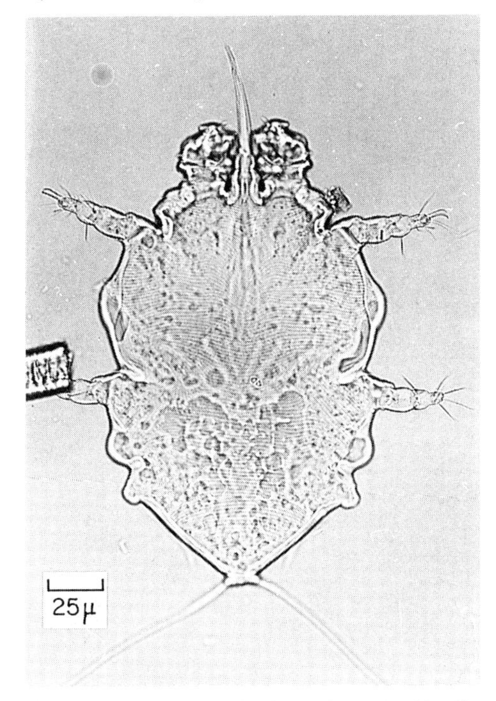

Fig. 2.11 Larva of the mite, *Radfordia ensifera*, showing double claws on the second pair of legs.

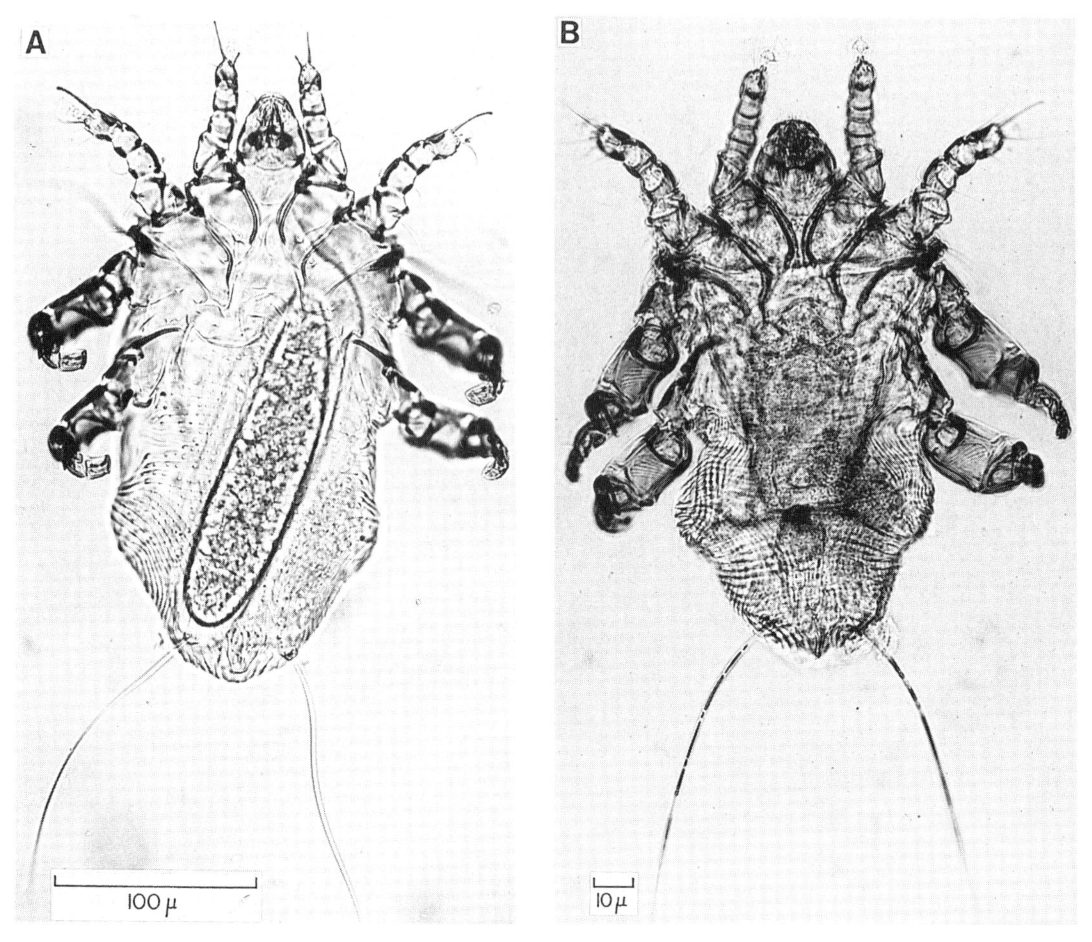

Fig. 2.12 *Myocoptes musculinus*, mange mite from a mouse. **A**: Adult female showing modified claws on the third and fourth pairs of legs. **B**: Adult male.

Fig. 2.13 A mouse with severe mange caused by both *Myobia musculi* and *Myocoptes musculinus*, together with secondary bacterial infection.

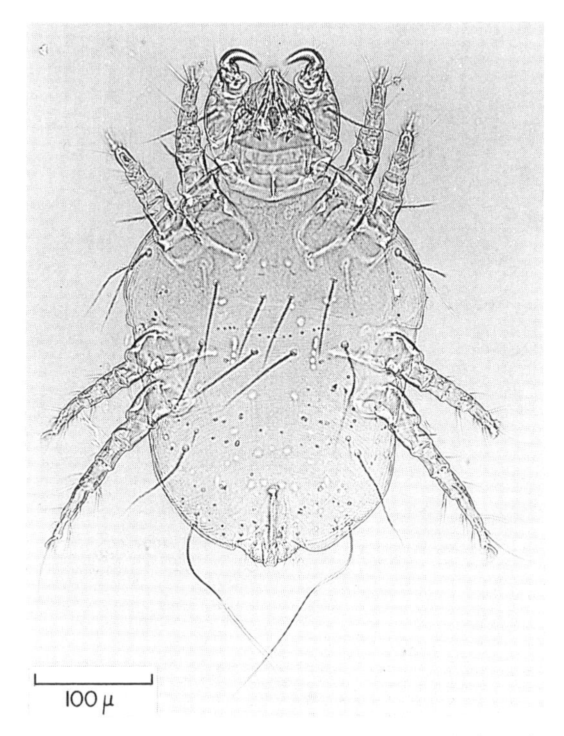

Fig. 2.14 *Cheyletiella parasitivorax*; adult female from a rabbit.

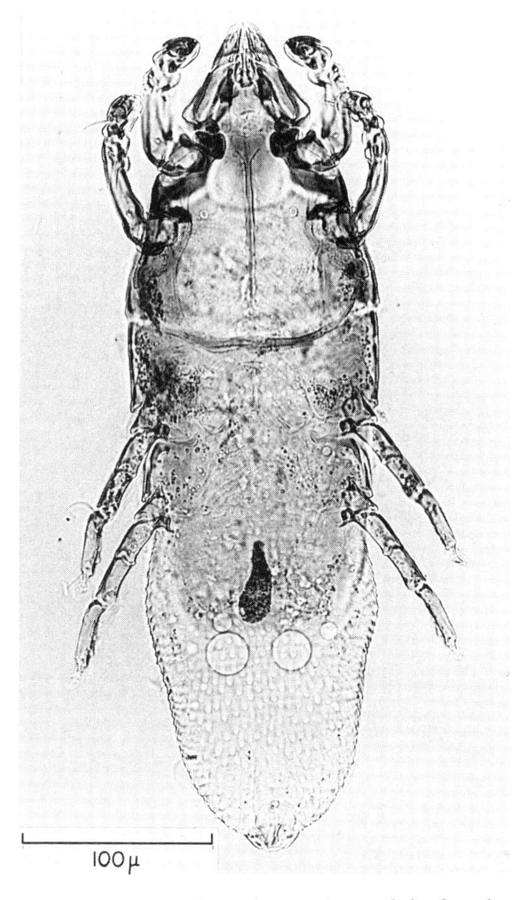

Fig. 2.16 *Chirodiscoides caviae*; adult female from the fur of a guinea pig

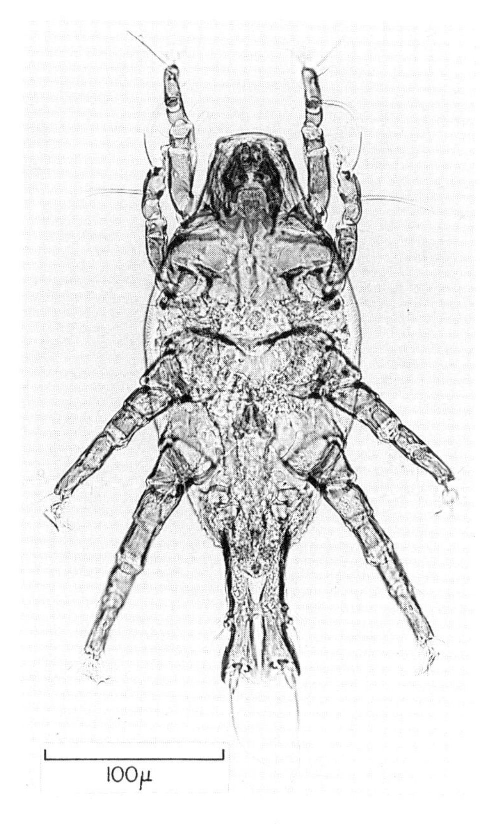

Fig. 2.15 *Listrophorus gibbus*; adult male from the fur of a rabbit.

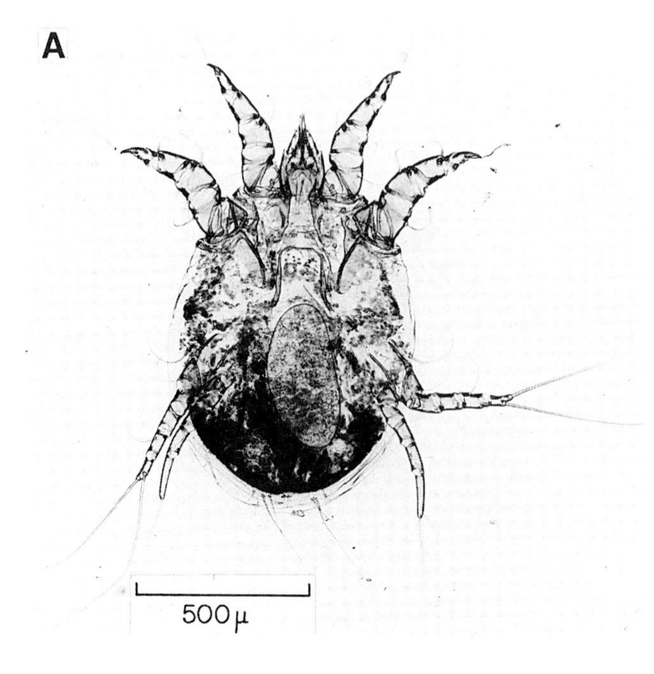

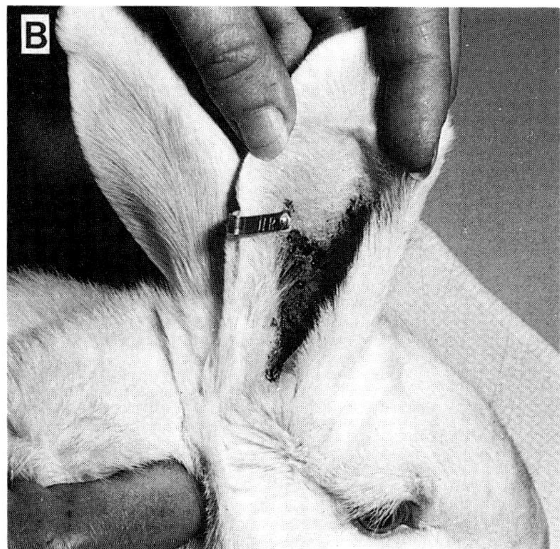

Fig. 2.17 *Psoroptes cuniculi*, the ear canker mite of rabbits. **A**: Adult female. **B**: Clinical lesions in the ear of the rabbit.

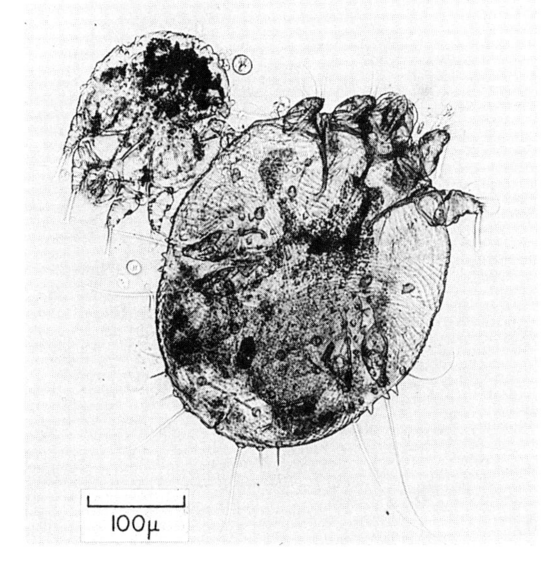

Fig. 2.18 *Sarcoptes scabiei*; adult and larva from a laboratory reared pig.

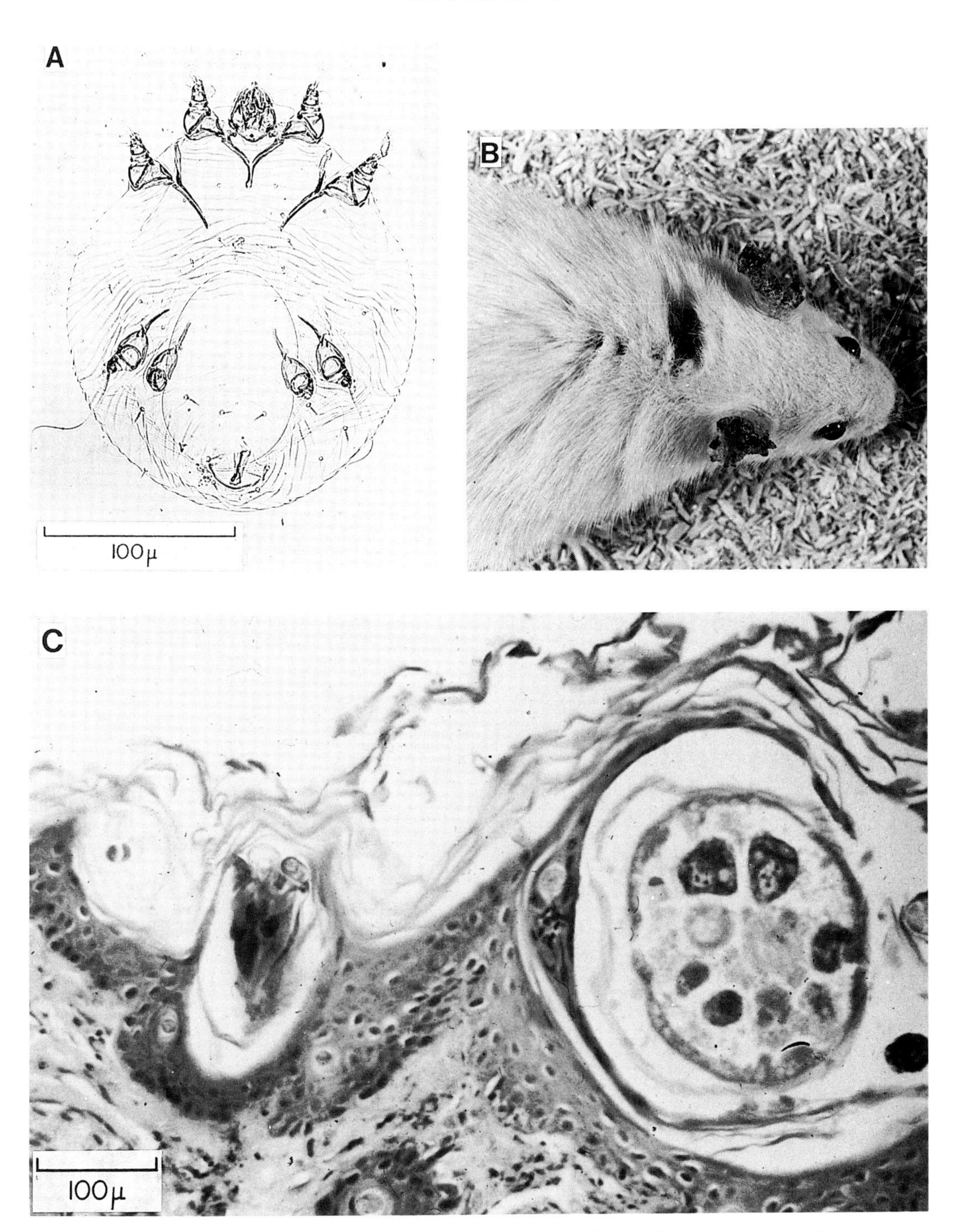

Fig. 2.19 *Notoedres muris*, a burrowing mite of rats. **A**: Adult female from a burrow in the skin at the tip of the ear. **B**: Rat showing typical lesions on the ears with spread of infection to the nose and the back of the head as a result of scratching. **C**: Section through the ear showing mites *in situ*.

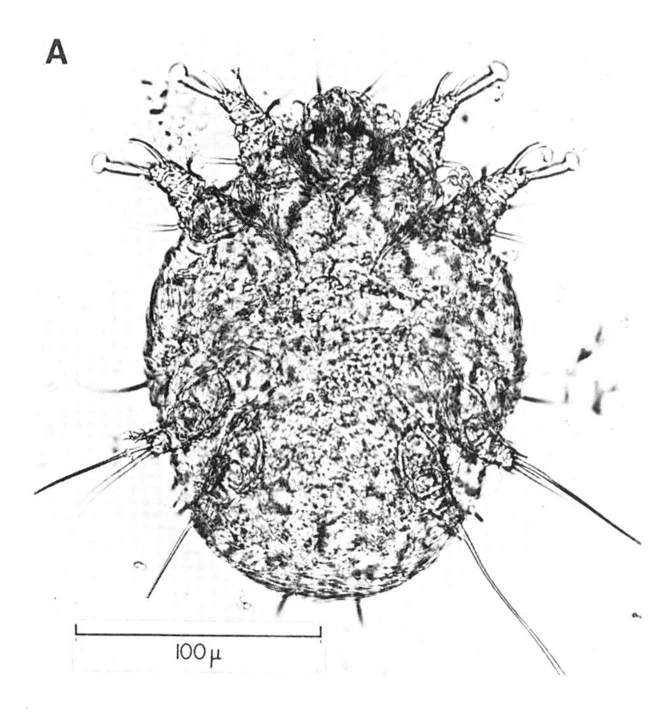

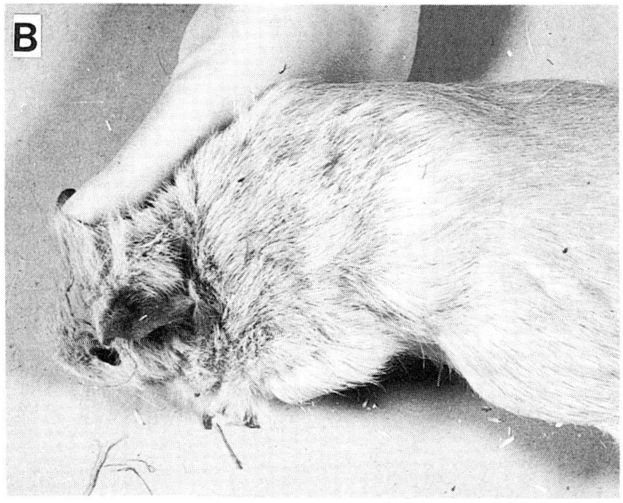

Fig. 2.20 *Trixacarus caviae*, a burrowing mite of guinea pigs. **A**: Adult from a skin scraping. **B**: Typical mange lesions.

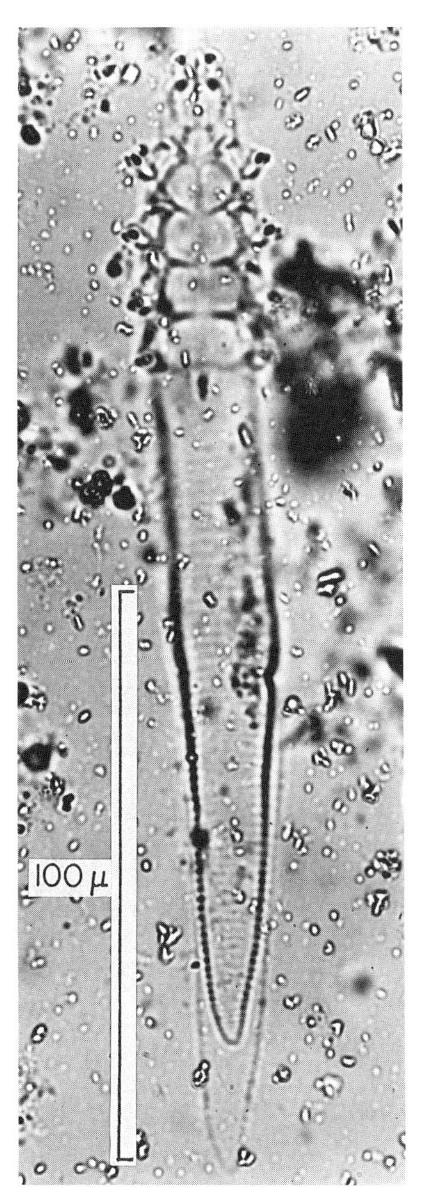

Fig. 2.21 *Demodex aurati*, follicle mite from the hamster.

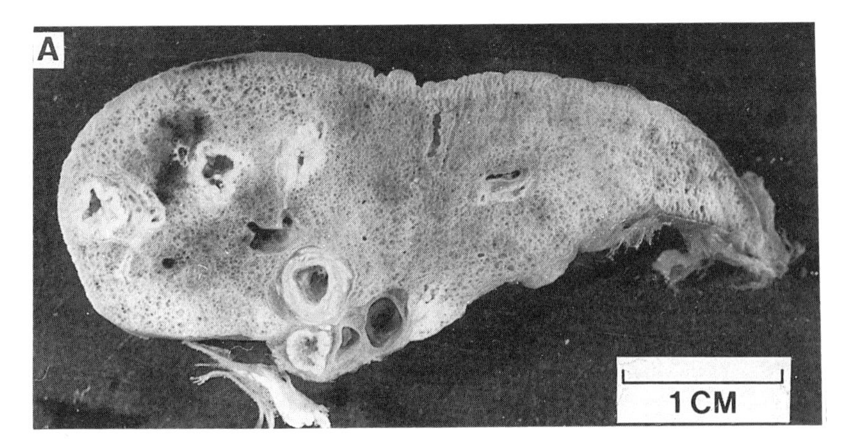

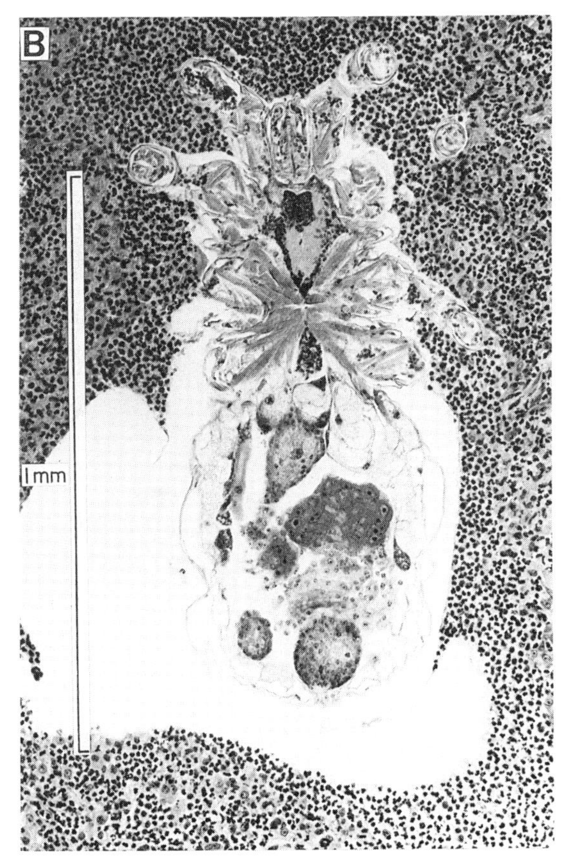

Fig. 2.22 *Pneumonyssus simicola*, lung mite of Old World monkeys. **A**: Effect of infection on the lung of a rhesus macaque. **B**: Section through an adult female in lung of baboon, showing strong cellular reaction.

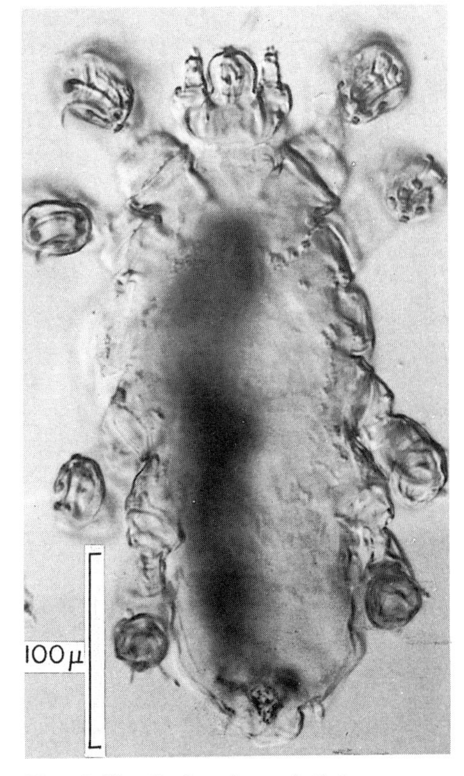

Fig. 2.23 *Spelorodens clethrionomys*, a mite found in the nasal cavities of hamsters.

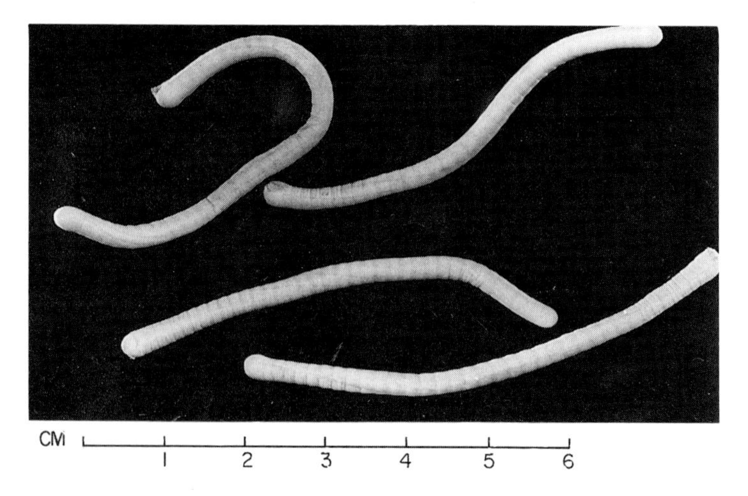

Fig. 2.24 *Porocephalus subulifer* adults.

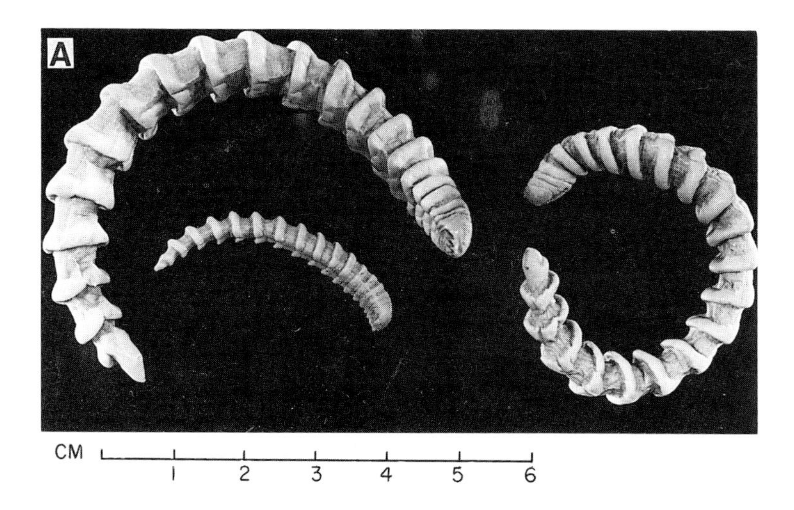

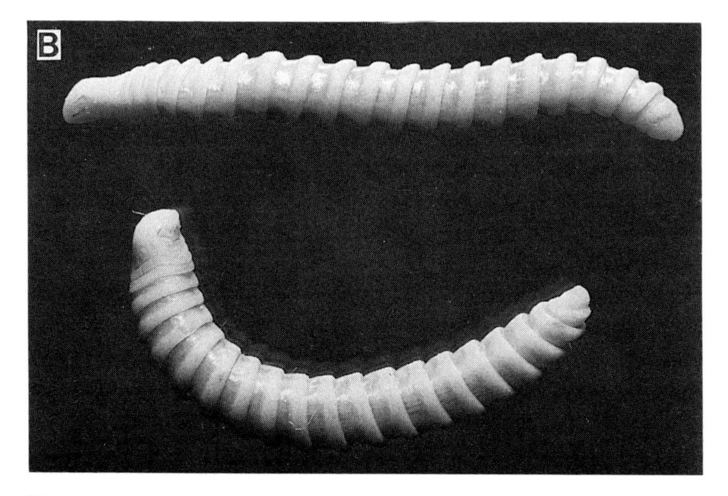

Fig. 2.25 *Armillifer armillatus*. **A**: Adults. **B**: Nymphs from the mesentery of a mongoose.

SECTION 3

ENDOPARASITES

A wide range of endoparasitic forms manage to survive the rigours of even the best animal husbandry. However, very great improvements in the standard of breeding and husbandry have reduced their variety and eliminated many of those with complicated life cycles. The endoparasites seen in the laboratory are usually restricted to a small number of species within each class or order and comprise those with the simplest and most direct life cycles. The host-parasite list is long and, as with the ectoparasites, the intensity of infection is favoured by the large number of susceptible hosts in a confined space, and the level of infection is usually much higher than that found in their wild counterparts. All the parasites discussed in the following section have been isolated either from wild-caught imported primates or from laboratory animal breeding colonies at some time over the last two decades.

PHYLUM PROTOZOA

The taxonomy of the Protozoa is in a serious state of flux at present. In the interests of clarity the classification followed here is taken from Honigberg *et al.* (1964). Those requiring a more recent review are referred to Margulis *et al.* (1989).

CLASS RHIZOPODA

Order Amoebida

Few, if any, of the many species of Rhizopoda found in the alimentary tract of laboratory animals cause serious clinical problems. The most common among hamsters, rats and mice is the non-pathogenic species *Entamoeba muris* (Fulton & Joyner 1948; Hsu 1982). It is located in the lumen of the large intestine, and is a large sluggish amoeba, its locomotion depending on the extrusion of pseudopodia. Food particles, such as bacteria, are engulfed by the pseudopodia and drawn into the cytoplasm where they circulate in food vacuoles during the process of digestion. The trophozoite measures around 25–30 μ and the large cysts are 15–20 μ in diameter.

The cyst is an important stage for the identification of different species as the features of the mature cysts can be very distinctive. In *E. muris* the nucleus has divided into eight small nuclei which can be seen clearly in the fresh unstained specimen and also in iodine-stained specimens together with splinter-like chromatoid bodies and a discrete glycogen granule (Fig. 3.1, page 47). Before encysting, the organism rounds up and all the food vacuoles are eliminated. There is a slight shrinkage in size and the delicate cyst wall is formed around the organism to protect it from the rigours of the outside world. Although unable to withstand drying, the cyst will protect the organism from a range of temperatures and can withstand a variety of dilute disinfectants (personal observation). Division into eight nuclei occurs early in the development of the cyst, division of the cytoplasm follows and eight individuals emerge at excystation in the small intestine of the new host. Once released into the lumen of the gut they feed and grow, multiplying by means of binary fission.

E. muris is often found in large numbers in the lumen of the caecum but it has never been implicated in any sort of enteric disorder. Similar non-pathogenic species are commonly found in the caecum of the guinea pig (*E. caviae*) and the rabbit (*E. cuniculi*) (Kheisin 1938). These and the non-pathogenic species found in man and primates (*E. coli*) are all very similar (Poindexter 1942). They are all large, lumen-dwelling forms with an eight-nucleated cyst roughly 10–30 μ in diameter and in no instance have they been associated with any pathological condition (Levine 1961).

In contrast, *E. histolytica*, a common parasite of man in the tropics, is known to exist in a variety of invasive forms. The degree of pathogenicity can vary between a non-clinical carrier state through to varying degrees of enteric disturbance, dysentery and, in some cases, invasion of other organs such as the liver where large sterile abscesses may be formed (Martinez-Palermo 1987). A number of factors can influence the virulence of this parasite. Sargeaunt & Williams (1978, 1979) discovered that populations of amoebae could be separated by means of their isoenzyme patterns and they described 20 of these populations which they called zymodemes. They found that invasive and non-invasive forms apparently always fell into specific categories so that each zymodeme could be graded according to virulence.

This is an important discovery, and if the zymodeme system is a stable one it will have considerable impact on the treatment of the infection worldwide (Sargeaunt 1987). However, certain bacteria have long been known to have an influence on the invasiveness of amoebae and the strict segregation of strains into 'invasive' and 'non-invasive' is not possible if all strains have at least a degree of inbuilt ability to invade. Recent work (Mirelman *et al.* 1986; Mirelman 1987*a*, *b*) has indicated that, at least *in vitro*, bacteria are able to alter the zymodeme from non-invasive to invasive. This, too, is an important discovery and if proven *in vivo* it could do much to explain some of the varying responses observed between individuals infected with the same zymodeme and which were not thought to be attributable to immune response.

Laboratory rodents are not natural hosts to *E. histolytica* but are known to be susceptible to infection with human *E. histolytica* trophozoites inoculated intracaecally. The rat (Neal & Harris 1977) and hamster (Neal & Vincent 1955, 1956; Williams 1959; Jurumilinta & Maegraith 1961) have long been the animal models of choice for research into caecal ulceration and liver abscess respectively. The mouse has also been found to be a good model (Neal & Harris 1975, 1977; Owen 1985) and some inbred strains of mice have shown a tendency to produce spontaneous liver abscess with certain strains of the parasite (Owen 1987). Despite this, there have not been any observations of natural cross-infections from humans to rodents in the wild or under laboratory conditions.

There have, however, been many reports of *E. histolytica* in laboratory primates, particularly in those where capture and holding have been close to human habitation. In man this is a very serious pathogen but in primates less so. There is, however, considerable variation in susceptibility between species—some New World species being sensitive to infection, some showing no clinical signs (Johnson 1941; Miller & Bray 1966). Advanced clinical lesions have been described in a range of different species (Bostrom *et al.* 1968), in spider monkeys (Eichorn & Gallagher 1916), in chimpanzees (Suldey 1924), and in orang-utans (Patten 1939). There are no reports of Old World monkeys developing dysenteric amoebiasis except in the case of dietary deficiency (McCarrison 1920). In the healthy host even experimental administration of pathogenic human strains failed to initiate infection (Hegner 1934; Miller 1952). However, in all cases where infection is established its course is similar to that seen in humans. Invasion is effected by an attack on the mucosal wall, the endotoxin secreted by the parasite effectively loosening the cells which are then phagocytosed by the amoeba. The invading trophozoite is markedly different from the lumen-dwelling form, being larger and much more active, and the food vacuoles usually contain red blood cells. Invasion of the mucosa is followed by an erosion of the mucosal wall by lytic enzymes and the formation of depressed lesions with raised edges. Sloughing of the mucosal wall and multiple lesions of this type can result in the destruction of large areas of the colon, and death may result.

The name *E. chattoni* is given to a common amoeba parasite of primates which may have been confused at times with immature *E. histolytica* cysts because of their similar size. They measure 10–18 μ. *E. chattoni* (or *E. polecki*; Kessel & Johnstone 1949), however, has a uninucleate cyst and in a formed stool should be readily distinguishable from the four nucleated cysts of *E. histolytica*. When stained with iodine the large nucleus of *E. chattoni* can often be seen occupying about a quarter of the cyst and is often pushed to one side by the pressure of numerous small glycogen granules. Though the majority of cysts contain a single nucleus, rarely there are two, and the chromidial bars are small and distinctive. Unlike *E. histolytica*, this is a non-pathogenic species, the trophozoite living in the lumen of the colon and the contents of the food vacuoles containing bacteria and yeasts.

E. hartmanni is a very small species common in primates. The trophozoite measures 6–12 μ and is almost impossible to distinguish from other small non-pathogenic amoebae of the colon such as *Dientamoeba fragilis* and *Endolimax nana* in wet preparation. The cysts are round or oval in shape and measure 4–10 μ. The four small nuclei will be seen on staining to have an eccentric karyosome and the chromidial bars have blunt ends similar to those of *E. histolytica*.

The cysts of *Iodamoeba buetschlii* are very irregular in shape, varying from round to oval or almost triangular and measuring 6–15 μ. There are rich glycogen masses within the cyst and these show up prominently in histologically stained specimens as a large colourless area but dark brown with iodine stain. The nucleus shows up as a large irregular karyosome with fine fibrils extending to the wall of the nucleus and with a clear area of surrounding cytoplasm (Wenrich 1937). There is usually only one nucleus per cyst. It is a very common non-pathogenic lumen-dwelling amoeba found in a wide range of Old and New World primates.

The mode of transmission of *Dientamoeba fragilis* is still unknown. The trophozoite can be seen in sections of the colon and appendix. It measures 5–15 μ in diameter when rounded and most have two nuclei. In man this amoeba has been known to produce a mucoid diarrhoea, but it has not been isolated as a natural infection from primates although they are known to be susceptible to experimental infection. It is thought that the course of infection when established is very similar to that in man (Burrows & Klink 1955).

Endolimax nana is another small, non-pathogenic amoeba found in the caecum and colon of Old and New World primates, and in some cases in the vagina (Hegner & Chu 1930). The trophozoite is small and sluggish, measuring 7–12 μ; the oval cyst is 8–12 × 5–7 μ in size and usually contains four nuclei, sometimes only two. The chromatoidal bodies and the nuclear structure of all these very small cysts are always difficult to observe in wet preparations but the small size and distinctive oval shape of the cyst is an aid to identification (Fig. 3.2, page 47).

Entamoeba gingivalis is an inhabitant of the mouth and the trophozoites can be isolated by scraping around the teeth. It is a common parasite of man, dogs, cats, rhesus monkeys and baboons as well as the great apes. It appears to increase substantially in numbers where there is concurrent gum disease and tooth decay, but there is no firm evidence that it initiates gum disease. No cysts have been isolated from this species (Burrows 1965). The small, sluggish trophozoite measures 8–25 μ and structurally is very similar to a lumen-dwelling *E. histolytica*, but the food vacuoles often contain white cells as well as bacteria, yeasts and other detritus. There have been no reports of a similar parasite in the mouths of rodents.

CLASS MASTIGOPHORA

The majority of the flagellate parasites of laboratory animals, like the amoebae, are non-pathogenic. There is, however, a much wider range of representatives, particularly in strict herbivores such as the guinea pig.

Order Trichomonadida

Trichomonas sp. (Fig. 3.3, page 47) may be found in the caecum of all rodents and primates considered here, but not lagomorphs. *T. muris*, found in the caecum of the mouse, is a large boat-shaped species with a stiff rod-like axostyle. Three anterior flagellae are directed forwards and a fourth is directed backwards and is attached along its length to the organism by means of an undulating membrane. The rippling movement of this membrane gives the organism its characteristic rolling and spinning movement. The trophozoite measures $16–26 \times 10–14\ \mu$ and has an anterior vesicular nucleus. Behind this is the source of the anterior and posterior flagella and the fibrillar costa, the blepharoplast, best seen in fixed smears by using Giemsa stain. Multiplication is by simple binary fission and no true cyst has been observed.

Despite the apparent lack of a protective cyst, various species of *Trichomonas* commonly found in laboratory rodents are often the first organisms to breech the barriers in specific pathogen-free (SPF) animal units established from caesarean-derived stock. Other non-pathogenic species such as *T. minuta*, a small oval species measuring $4–9 \times 2–5\ \mu$, are also common contaminants. Although non-pathogenic themselves, the presence of these flagellates can act as an early warning of a potentially more serious breakdown in the barrier.

Trichomitus wenyoni in rhesus monkeys and baboons (Wenrich & Nie 1949), *Tritrichomonas caviae* in the guinea pig (Nie 1950), and *T. criceti* in the caecum and colon of the hamster are all very similar in appearance and measure $4–16 \times 2.5–5\ \mu$ (Wenrich & Saxe 1950).

The guinea pig seems able to tolerate a wide range of caecal flagellates often in very high numbers with no apparent clinical signs (Nie 1950). *Hexamastix*, *Chilomitus*, *Enteromonas* and *Retortamonas* sp. are found in very large numbers and these vary individually in the number of anterior flagella and the presence or absence of a cytostome or cysts. They are difficult to distinguish from each other in wet preparations, so coverslip preparations should be made and stained with Giemsa (*see* Appendix B).

Chilomastix sp. are larger flagellates, measuring $6–20 \times 3–10\ \mu$. Typically they have a firm pellicle but no marked axostyle; three anterior flagella are directed forward and a fourth flagellum can be seen flickering within a deep cytostomal groove. It has a distinct lemon-shaped cyst (Fig. 3.4, page 47) which measures $6–10 \times 3–6\ \mu$ and is colourless in wet preparations, but when iodine-stained the large oval nucleus and the fibrils and cytostome flagella can be seen (Burrows 1965). The trophozoites move in a spinning spiralling motion and multiplication is by means of binary fission. *C. mesnili* is a common parasite of man and great apes; *C. intestinalis* and *C. wenyoni* both occur in the caecum of the guinea pig; *C. battencourti* in the caecum of the hamster (Wantland 1955), the mouse (Bonfante *et al.* 1961) and the Norway rat (Gieman 1935); and *C. tarsii* in the tarsier (Porter 1966). All these species are thought to be non-pathogenic although they are invariably found in company with a range of other protozoa.

Only one species of *Trichomonas* has been found in the mouth, that is *T. tenax*, a very small non-pathogenic species (6–12 × 3–7 μ) found in primates and man (Wenrich 1947). *T. vaginalis*, on the other hand, is a very large species, found in the vagina of primates and man although there is little evidence of pathogenicity. A species similar to *T. bovis*, which causes abortion in cattle, was isolated from a tarsier where it was associated with fetal abortion (Hill *et al.* 1952).

There are, however, two flagellate species, inhabitants of the small intestine, that have been found to be pathogenic under certain circumstances. These are *Spironucleus (Hexamita) muris* (Fig. 3.5, page 48) and *Giardia muris* (Fig. 3.6, page 48). *S. muris* is a small, pyriform organism measuring 7–9 × 2–3 μ and is bilaterally symmetrical. The two nuclei are located at the anterior end and six long flagella emerge at a point just above. Two further flagella are directed backwards along the length of the body and trail posteriorly alongside short, stubby, twin axostyles. Reproduction is by binary fission and distinctive ovoid cysts are formed which measure 7.4 × 4.0 μ (Kunstyr 1977). In fresh wet preparations made with normal saline the organisms can be seen moving in short rapid bursts, always in a straight line rather like a torpedo.

S. muris is a common parasite, readily transmissible within a range of laboratory rodents in which it is an inhabitant of the mid to lower small intestine; it is only associated with clinical signs in certain strains of mouse (Meshorer 1969; Sebesteny 1969, 1974, 1979), although changes in macrophage metabolism have been observed (Keast & Chesterman 1972). The signs of the disease usually first appear in young stock and consist of rapid weight loss and diarrhoea, often resulting in death. Post-mortem examination often reveals a 'gassy gut' syndrome and sections of the gut wall reveal massive numbers of the organisms in the crypts and mucosal lesions. Older mice suffer a chronic condition where weight loss is less severe and they only rarely have diarrhoea. Few adults die of the infection and the lesions in the small intestine are confined to the duodenum. A similar species has been found in primates: *Hexamita pitheci* in the rhesus monkey in South America (da Cunha & Muniz 1929) and in rhesus monkeys and chimpanzees in North America (Wenrich 1933; Hegner 1934).

Giardia lamblia is a common parasite of man, in whom certain strains are known to cause enteritis and may be zoonotic (Faubert 1988; Bemmich & Erlandsen 1988), crossing a range of animal species. It has also been found in a range of primates and the pig, in which it is thought to be the cause of diarrhoea but insufficient data have been collected to confirm this. This bilaterally symmetrical organism, with the nucleus situated at the anterior end, has a strange pear-like shape with the anterior ventral surface broadly concave and shaped like a sucking disk. As with *S. muris* there are two axostyles and eight flagella emerging in pairs around the rim of the disk. Multiplication is by binary fission and cysts are produced which are voided with the stool. The cysts measure 8–12 × 10 μ, are oval in shape, and, when mature, the nuclei and axostyles of the young *Giardia* can be seen within the cyst in wet preparations although staining with iodine will show them better. *Giardia* apply themselves closely to the surface of the gut epithelial cells where, in very heavy infections, they may carpet large areas of the upper small intestinal surface and eventually cause severe digestive disturbance. Species of *Giardia* have been described in rats, outbred mice and hamsters (Filice 1952; Levine 1961) and are often found in very large numbers in adult animals with no apparent ill effects (personal observation). However, there is a pot-bellied condition often seen in weanling hamsters which may be associated with very high numbers of both *G. muris* and *S. muris*. This condition usually resolves itself spontaneously in a few weeks, the animal's condition improving and the swollen abdomen receding in association with a reduction in the number of both types of flagellate.

G. caviae of the guinea pig can sometimes be found in the duodenum but, like *G. duodenalis* of the rabbit, there have been no reports of clinical disease (Hegner 1923).

Suborder Trypanosomatina

Trypanosomes are flagellates found inhabiting the bloodstream of mammals. The trophozoites are usually long slender crescents, very flexible and active, having a single long flagellum arising posteriorly and directed forwards along the length of the body. The flagellum is always attached by an undulating membrane and may continue forward as a free flagellum. In some species there are only long slender forms with a free flagellum, in others only short stumpy forms with no free flagellum, and in other species both forms coexist. They are easily recognized in wet blood preparations by their characteristic lashing movement.

In the wild, *Trypanosoma lewisi* (Fig. 3.7, page 49) is a common blood parasite of rats with a range of rodent fleas that can act as vectors. It is monomorphic, being a long, slender parasite measuring 25 μ in length. When first introduced into a new host by the vector, it undergoes a ten-day period of multiplication. These large, broad trypanosomes divide rapidly by longitudinal fission. The next phase consists of slender non-reproductive forms which remain in the blood through the rest of the infection, often lasting several months before the immune system of the host eliminates them. The flea is infected by ingesting the circulating trypanosomes in a blood meal. The organisms enter the cells of the stomach wall of the flea where they undergo a period of development and multiplication. They then progress along the alimentary tract, continuing their phase of multiplication as they go, and emerge in the excreta of the flea as small forms approximately five days after infection. Infection of the mammalian host is a result of the wound made by the flea during feeding being contaminated by the trypanosoma. The morphology of the trypanosome can be clearly seen in thin blood films dried, fixed and stained with Giemsa. Near the point of attachment of the flagellum is the large dark kinetoplast and the large central nucleus. Although common in rats in the wild, this parasite has not been reported in laboratory colonies because of the absence of the vector. It is generally thought to be non-pathogenic and the condition is soon resolved.

Trypanosomes are common pathogens of man throughout Africa and South America. In Africa transmission is by means of biting flies, usually *Glossina* species (tsetse fly). In monkeys trypanosomes are not associated with natural disease, but they are known to be susceptible to the species causing disease in man because experimental infections with *T. gambiense* and *T. rhodesiense* result in heavy, chronic infections which may be fatal.

Closely related are the forms of *Leishmania*, which are tissue-invading flagellates usually found within the reticuloendothelial cells of the host. They consist of a small, round body measuring 1.5–4 μ in diameter with nucleus and kinetoplast but no flagellum. They multiply and develop in the cells of the reticuloendothelial system and are transmitted by small biting flies, *Phlebotomus* sp. Within the insect host the organisms develop intracellularly as amastigotes in which there are no signs of a flagellum. They then develop a flagellum, becoming firstly a promastigote and then an active leptomonad form. Only rarely have natural infections been found in primates (Wenyon 1926; Baker 1972) but macaques are sufficiently sensitive to infection to be widely used in the study of human leishmaniasis.

CLASS CILIATA

The guinea pig is the only laboratory rodent species likely to be found harbouring enteric ciliates. The most commonly and easily observed is a species of *Balantidium*, *B. caviae*, found in the lumen of the caecum often in large numbers (Fig. 3.8A, page 49). It is a large oval organism, varying in size from 50–120 μ in length by 45–80 μ in width. The whole of the flexible pellicle is covered with fine cilia arranged longitudinally and finishing around an oral groove or cytostome. Typically there are two nuclei, a large spherical or kidney-shaped macronucleus and close alongside it a small spherical micronucleus. At the posterior end of the body the anal opening or cytopyge marks the exit of an excretory duct which leads from a contractile vacuole which controls osmoregulation. During the process of feeding the cilia waft food particles along the cytostome and they are engulfed into food vacuoles at the end of the groove.

Reproduction is usually achieved by means of transverse binary fission, and a complex process known as conjugation may precede fission. When this occurs two individuals fuse together at the oral surface and proceed to undergo nuclear rearrangement and exchange where a (daughter) nucleus from each individual passes across and fuses with the remaining nucleus in the other. They separate and proceed to divide by fission.

The resistant stage of the parasite is a large dark-coloured, thick-walled cyst which measures 40–60 μ. The most prominent structure in the cyst is the macronucleus. The organism is easily identified in wet preparations by means of its large size and by its rapid gliding motion.

Although *B. caviae* is common in the guinea pig and often present in large numbers, there are no reports of any clinical condition associated with the infecion, as there have been in domestic pigs and in man. Numerous species of *Balantidium*, e.g. *B. coli* (Fig. 3.8B, page 49), have also been recorded in non-human primates but pathological conditions are more common in apes than in monkeys. Clinical signs in apes are generally similar to those found in amoeba infection and the loss of appetite results in weight loss, diarrhoea and rectal prolapse. The organisms invading the mucosa may be seen histologically in groups or singly with or without associated lymphocytic infiltration. They may cause extensive necrosis in the submucosa.

Another common ciliate of conventional guinea pigs is *Cyathodinium piriforme* (Lucas 1932*a*, *b*; da Cunha & de Freitas 1940). This is a cone-shaped organism being broad anteriorly and tapering to a posterior tip which is bent dorsally. It is lighter in colour than *Balantidium* with a more fragile appearance. The cilia are confined to the anterior region and, unlike other holotrichous ciliates, which have connecting fibrils responsible for their smooth synchronous movement, in this species each cilium moves independently of the others. The ellipsoidal macronucleus is central in position with the micronucleus closely applied to the side of it. Feeding is typically by the wafting of food particles into the peristome by the cilia and the production of food vacuoles. Cysts are rare. Nie (1950) described cysts in one animal as oval and measuring 17–24 × 12–17 μ. The active trophozoites are best seen in very fresh wet preparations where their characteristic heart-shape and gliding movements are easily identified. They appear less active than *B. caviae* and soon cease any activity shortly after the preparation is made. There have never been any reports of a clinical condition resulting from their presence.

Other non-invasive species from different ciliate families have been found in non-human primates: *Taliaferria clarki* in spider monkeys (Hegner & Rees 1933) and *Troglodytella* sp. in chimpanzees and gorillas (Brumpt & Joyeaux 1912; Reichenow 1920; Swezey 1932). In only one instance has diarrhoea and death been considered to be associated with *Troglodytella* infection in gorillas (Curasson 1929), though there appears to be some room for doubt that it was the primary initiator of the condition (Burrows 1972).

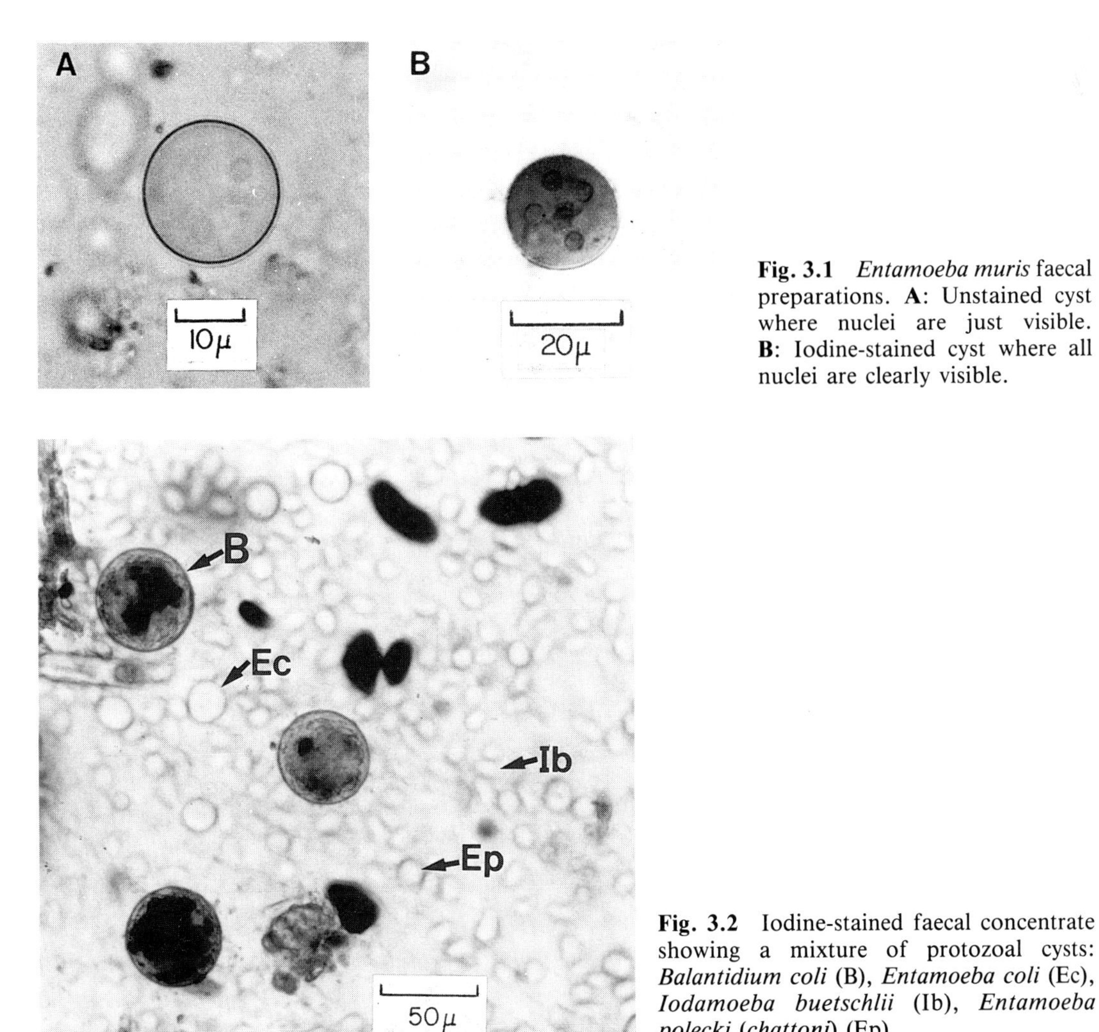

Fig. 3.1 *Entamoeba muris* faecal preparations. **A**: Unstained cyst where nuclei are just visible. **B**: Iodine-stained cyst where all nuclei are clearly visible.

Fig. 3.2 Iodine-stained faecal concentrate showing a mixture of protozoal cysts: *Balantidium coli* (B), *Entamoeba coli* (Ec), *Iodamoeba buetschlii* (Ib), *Entamoeba polecki* (*chattoni*) (Ep).

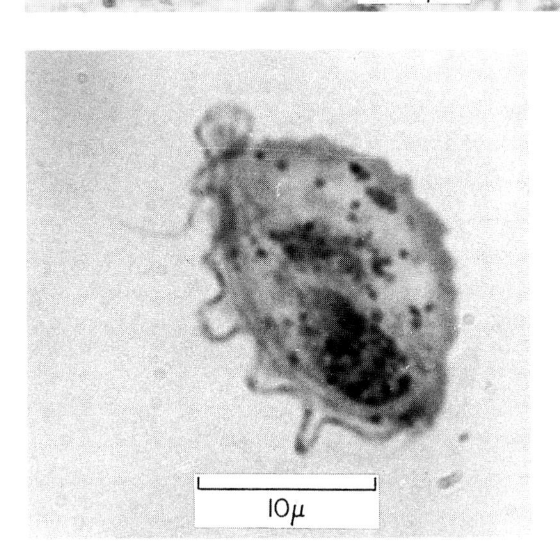

Fig. 3.3 Common *Trichomonas* sp. from the caecum of a hamster.

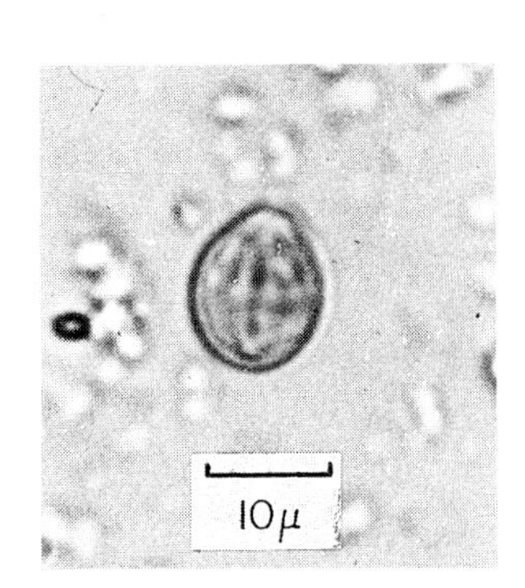

Fig. 3.4 Lemon-shaped cyst of *Chilomastix mesnili* from a baboon.

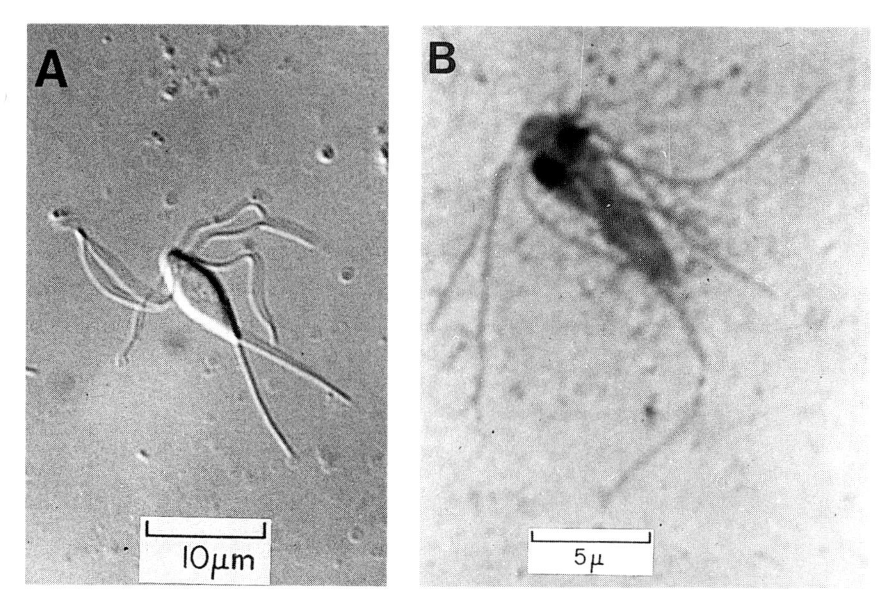

Fig. 3.5 *Spironucleus (Hexamita) muris*. **A**: Living trophozoite (phase contrast; kindly supplied by Dr Ivo Kunstyr, Medizinische Hochschule, Hannover); **B**: Giemsa stained trophozoite (kindly supplied by Mr Andor Sebesteny, Imperial Cancer Research Fund, Potters Bar).

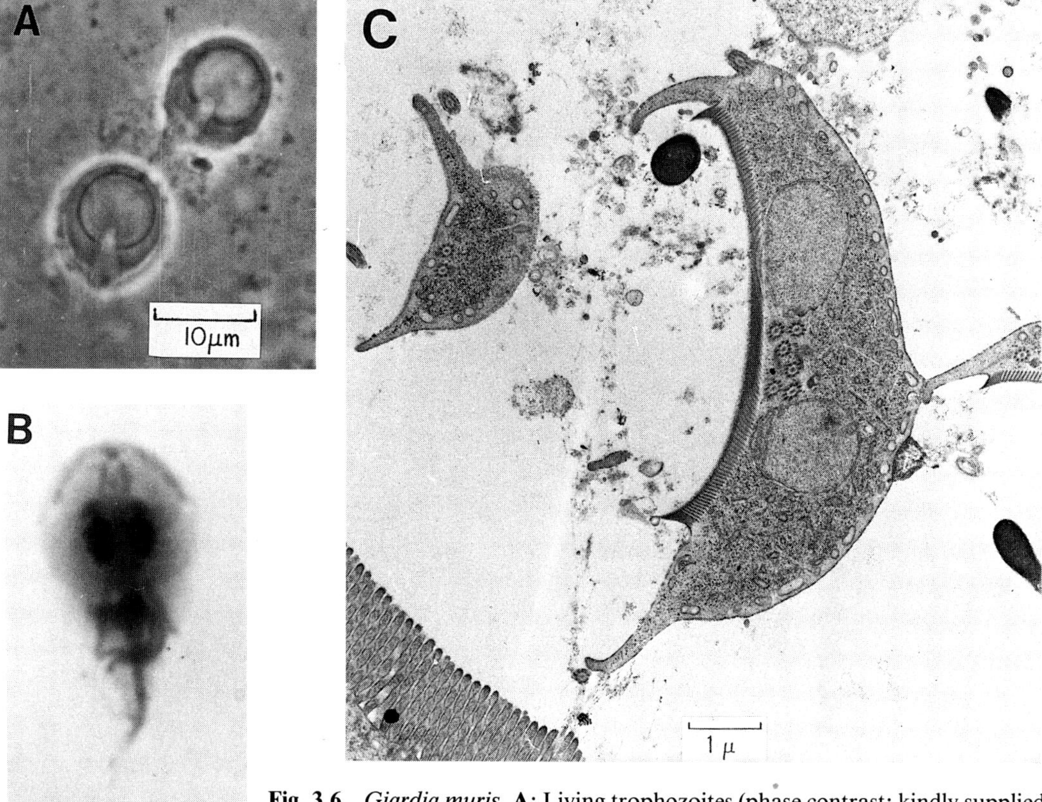

Fig. 3.6 *Giardia muris*. **A**: Living trophozoites (phase contrast; kindly supplied by Dr Ivo Kunstyr, Medizinische Hochschule, Hannover). **B**: Giemsa stained trophozoite from a wet preparation. **C**: Electron micrograph of organism in duodenum showing concave ventral surface which adheres to mucosal wall.

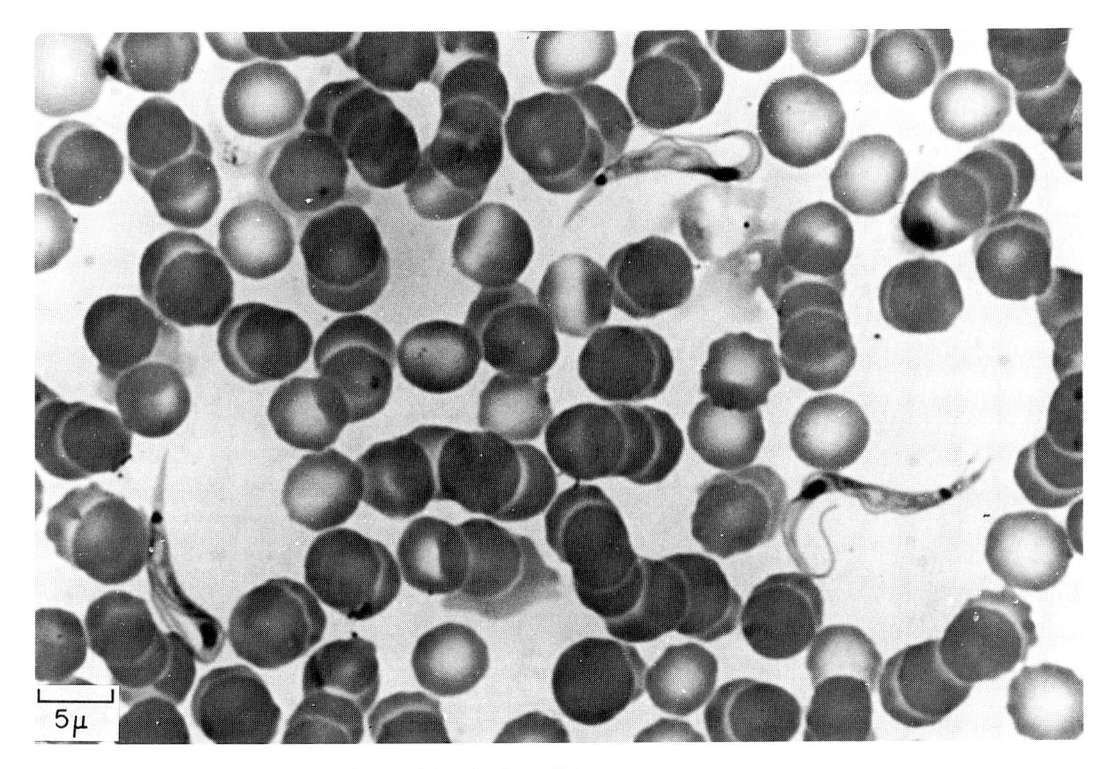

Fig. 3.7 *Trypanosoma lewisi* in the blood of a wild rat.

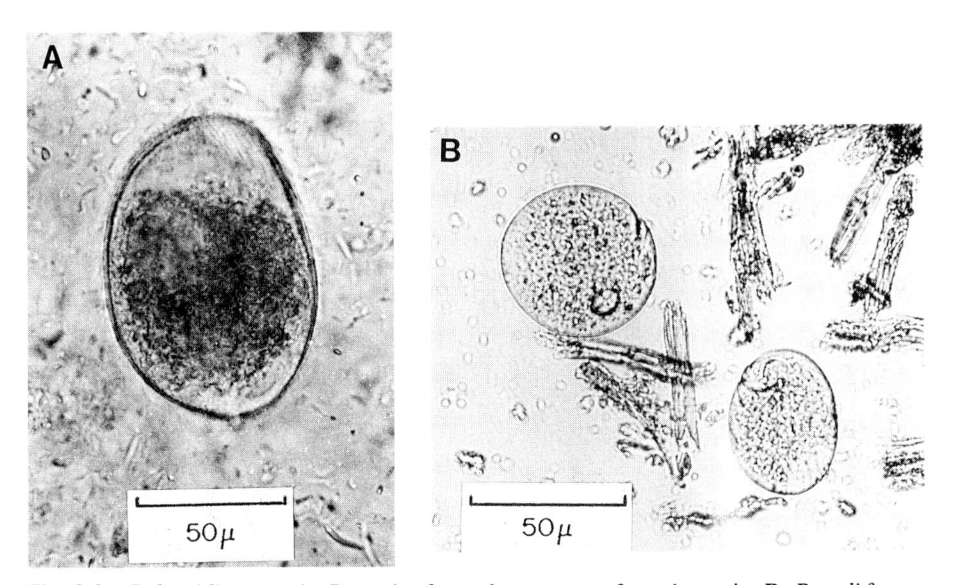

Fig. 3.8 *Balantidium* sp. **A**: *B. caviae* from the caecum of a guinea pig. **B**: *B. coli* from stool of *Macaca mulatta* (kindly supplied by Dr J Remfry).

CLASS SPOROZOASIDA: ORDER EUCOCCIDIORIDA

This particular group, the Sporozoa, contains the largest number and variety of protozoan pathogens of mammals, having within it the coccidia, haemosporidia and microsporidia (Levine & Ivens 1965; Levine 1973).

Coccidia

The coccidia of greatest concern to the laboratory animal colony are organisms belonging to the suborder Eimeriorina. All rodents are natural hosts to more than one species of *Eimeria* in the wild, but only rarely are they found in the laboratory if the level of hygiene is good. Only in lagomorphs, where as many as 14 species of *Eimeria* have been described, are these parasites a serious problem.

Eimeria sp. (Fig. 3.9, page 63) are parasites of the epithelial cells of the gut mucosa and those lining certain ducts. The resistant cyst stage is known as an oocyst and in all intestinal species it is voided in the faeces. The size and general shape vary between species but generally the course of development is similar. The oocyst when voided is undifferentiated and contains a single cell or sporont which requires oxygen if it is to develop further (Durr & Pellérdy 1969). Under suitable conditions division occurs over a period of days, depending on the species, into four sporocysts each of which contains two sporozoites. The sporocysts are clear envelopes in which the comma-shaped sporozoites can be seen, tapering at one end with a distinctive clear refractile granule at the other. The granule is proteinaceous and is a diagnostic feature of the sporozoite and of the first generation schizonts. Once ingested, the oocyst is subjected to digestive enzymes which stimulate the hatching of the sporozoites from the sporocysts and then, in the duodenum, from the oocyst.

Eimeria are not only host-specific, they are also site-specific and invade specific epithelial cells along the length of the small intestine, caecum or colon. The location and the cells they invade together with the depth to which they penetrate are important factors influencing the severity of the disease. Once within the chosen cell, the sporozoite rounds up and over a period of hours or days forms a trophozoite, the nucleus of which divides, then the cytoplasm, forming numerous long, slender, sickle-shaped bodies known as merozoites in what is now known as a schizont. These emerge from the cell when mature and proceed to find another cell where this asexual multiplication is repeated. The number of asexual stages as well as their site and number of merozoites vary with the species, but finally a sexual generation is produced when the merozoites entering the cells develop into male and female gamonts. The microgametocytes (male) produce a large number of flagellated microgametes, while the macrogametes (female) enlarge filling the cell. Once fertilized by a microgamete this becomes a zygote, forming a firm wall and eventually being shed into the lumen of the gut and voided to the exterior as an unsporulated oocyst. The oocyst in a great many species is very resistant to changes in the environment and under suitable conditions can survive for several years. A pure strain of *E. stiedae* from the rabbit, kept by the author in 2.5% potassium dichromate at 4 °C, hatched when digested six years later. They are, however, sensitive to bacterial contamination and cannot survive drying or dry heat.

Eimeria sp. in the rat
The rat has three common species in the wild, of which two are found in the small intestine, *E. nieschulzi* and *E. miyairii*; and one in the caecum, *E. separata*. The oocysts of

all three species found in the faeces are very distinctive and easily separated. *E. nieschulzi* is ellipsoidal or ovoid and the wall is smooth and colourless (Roudabush 1937). It measures 16–26 × 13–21 μ and has no micropyle (a thinning of the shell at the anterior end and often an important feature in identification) nor a residual body (a mass of cellular debris left over from sporulation and again a stable feature taken into account for identification). The oocyst of *E. miyairii* is spherical or sub-spherical with a thick rough wall, usually a dark yellowish-brown in colour, and measures 17–29 × 16–26 μ (Matubayasi 1938); it also lacks a micropyle and oocyst residual body. *E. separata* is the smallest species, with an ellipsoidal oocyst measuring 10–19 × 10–17 μ, and also has no micropyle or oocyst residual body. The wall is smooth and clear and usually colourless.

The prepatent period of *E. nieschulzi* lasts six days and consists of four asexual generations, the number of merozoites in each generation varying between 20–36 as in the first generation to 10–14 as in the second. Oocysts are released over a period of 4–7 days and if conditions are favourable will sporulate in two to three days. This is perhaps the most commonly found and most pathogenic of the three species, affecting mostly young animals in which heavy infections can result in death. Usually the profuse watery diarrhoea results in a general debility and dehydration. As the condition, like all *Eimeria* infections, is self-limiting, those animals that survive recover fairly rapidly and are usually immune. *E. miyairii* is generally similar in the number of asexual generations and the length of the prepatent period; however, the developing stages are found deep in the villi often below the epithelium. The pathogenicity of *E. miyairii* has not been thoroughly investigated, but the lower number of merozoites at each generation and the fact that fewer oocysts are produced could account for its apparently lower pathogenicity when compared to *E. nieschulzi* (Wright 1985). *E. miyairii* is much less commonly found in wild-caught animals and has not been reported in laboratory colonies. In contrast, *E. separata* is often found and is relatively non-pathogenic, even in heavy infections (Becker 1934). Its prepatent period is also six days and all the stages of development can be found in the caecum and colon.

Eimeria sp. in the mouse

Many species of *Eimeria* have been described in the mouse (Pellérdy 1965; Levine & Ivens 1965). Unfortunately many of the descriptions rely entirely on differences in the shape and size of the oocyst, with little or no further investigation. Two commonly found caecal species from wild-caught and laboratory specimens (*E. falciformis* and *E. pragensis*) have been isolated and cloned by the author. These, together with a small intestinal species, *E. vermiformis* (Ernst *et al.* 1971; Todd & Lepp 1971), will be considered here.

The true description of *E. falciformis* is somewhat confused, the name having been given to different species over the years. In this account the name *E. falciformis* is given to a caecal species which is non-pathogenic, has a maximum of four asexual generations, the final one at the same time as gametogony, and a prepatent period of 5 days only. Descriptions have been given by Cordero del Campillo (1959), Cerná (1962) and Haberkorn (1970).

The author's examination of the life cycle was carried out using a cloned strain of the parasite under strict gnotobiotic conditions (Fig. 3.10, pages 64–65). Histologically all stages of development could be seen taking place in cells on the sides and tips of the caecal glands, never at the base and always located above the cell nucleus. *E. falciformis* is fairly common, having profuse oocyst production and a sporulation time of 2–3 days. Pathogenicity is so low that only with doses of 50,000 oocysts or more was diarrhoea observed (personal observation).

On the other hand, the species named by Cerná & Senaud (1969) as *E. pragensis* is another common parasite in which development takes place deep in the crypts of the caecum and colon. This site, together with an increased prepatent period of eight days, not unexpectedly results in a much more pathogenic infection. There are at least five asexual generations and although there are no more than 20 merozoites at each generation their location results in a sloughing of the mucosa, particularly in the descending loop of the colon, in heavy infections. The clinical manifestations are typical of coccidiosis—profuse and often bloody diarrhoea, loss of appetite, wasting and weight loss and, in the case of *E. pragensis*, death can occur with any infection of over 1000 oocysts, or even fewer in very susceptible strains of mouse (personal observation).

There is no difference between *E. falciformis* and *E. pragensis* in the size and appearance of the oocyst except for the presence of a polar granule in the oocyst of *E. pragensis*. The correct naming of these species needs more attention. The author has selected the above names on the basis that the isolate named *E. falciformis* most resembles the descriptions given by Cordero del Campillo (1959) and Haberkorn (1970), although both their reports describe schizont stages in the ileum which were not found by the author in this species nor in that of *E. pragensis* described by Cerná & Senaud (1969).

The small intestinal species of the mouse, *E. vermiformis*, is relatively avirulent. It affects the lower two-thirds of the small intestine. Todd & Lepp (1971) considered it to be the species most likely to have been confused with *E. falciformis*, giving rise to the idea that it occupied both the small intestine and the caecum. The prepatent period of this species is seven days and two asexual stages have been observed, with gametogony commencing 5 days post-infection.

Eimeria sp. in the guinea pig

A single species, *E. caviae*, has been seen in the guinea pig. The oocyst is smooth, oval and light brown in colour and measures 13–26 × 12–23 μ with both an oocyst and sporocyst residual body but no micropyle or polar bodies. The developing stages are found in the colon, occasionally in the caecum where they invade the epithelial cells of the crypts mainly, but have been found throughout the mucosa. Schizonts, producing 12–32 merozoites, have been observed but the number of asexual generations is not known. The prepatent period lasts for 11–12 days (Henry 1932; Lapage 1940). The stages of development are usually restricted to the colon, where, in heavy infections, the lesions can be seen macroscopically through the serosal wall as pale white or yellow plaques together with petechial haemorrhages.

Pathogenicity is related to the conditions under which the animals are kept and the level of infection. Although supposedly non-pathogenic, it has been the author's experience that crowded floor pens, where oocysts are allowed to build up, invariably result in deaths among young stock.

Eimeria sp. in rabbits

A similar situation is to be found in keeping and rearing rabbits. In this case, however, there are around ten different species of *Eimeria* commonly found in rabbit houses, as well as many other named species and even varieties of some of the more common species that vary in their pathogenicity. As with coccidial infections in any other host, once an infection has run its course immunity to reinfection is usually very solid, but at no time is any cross-immunity built up between the species of parasite. All but one of the *Eimeria* species of the rabbit are found in the small intestine, caecum or colon, and most have a reasonably distinctive oocyst structure or colour (Fig. 3.11, pages 66–67).

One species however, *E. stiedae*, is an inhabitant of the epithelial cells of the bile ducts and is the cause of severe liver damage in rabbits (Fig. 3.12, page 68). When the oocyst hatches in the duodenum the sporozoites proceed to invade the wall and burrow into the mucosa. Once there they are carried to the liver either in the bloodstream or engulfed within a macrophage which travels via the mesenteric lymph nodes. The sporozoites therefore reach the liver in waves, those in the blood being detected as early as 12 hours post-infection and those carried by macrophages taking up to 96 hours and more. The mesenteric lymph node route reaches a peak at around 72 hours, when impression smears of the node will be seen to contain crowds of macrophages all with sporozoites easily detected by their distinctive refractile granules (Owen 1974). The target of the migrating sporozoites is the epithelial lining of the bile ducts. Once established, development commences in the same way as in the intestinal species with a series of asexual generations. Because of the staggered waves of infection it is difficult to determine the number of generations, although Pellérdy & Durr (1970) claimed to have found 9 different schizont stages. Most of the endogenous stages develop below the cell nucleus but also have been reported in liver perenchymal cells. The prepatent period lasts for 15–16 days and oocysts are found in the stool for at least 10–14 days after this.

Pathogenicity is variable: light infections may not produce any clinical signs but heavy infections inevitably result in anorexia, severe diarrhoea, distended abdomen and often death, especially in younger animals. At post-mortem examination the liver is usually very enlarged with a mottled whitish surface marking the distended bile ducts, and the gall bladder is vastly distended and packed with oocysts. Wet smears taken from the contents of the liver nodules will reveal the stages of the parasite, and histological sections of the dilated ducts show how the bile ducts have proliferated in an attempt to cope with the damage caused by the invading merozoites. The recovery rate from this infection is variable depending on the level of challenge and the age of the host. As always in coccidial infections, the most vulnerable are newly weaned animals, but those that recover have a very solid immunity.

Of the intestinal species in the rabbit the most pathogenic are *E. magna* and *E. irresidua*. The endogenous stages of the former are all situated below the epithelial cell nucleus and in the submucosa. All the developmental stages are restricted to the ileum and jejunum, but in heavy infections an overspill into the caecum has been observed. There are two asexual stages and oocysts appear in the faeces 7–8 days post-infection. Clinical symptoms of infection are characteristic in all intestinal species and consist of weight loss, anorexia, a profuse mucus diarrhoea and death. At post-mortem examination the affected portions of the gut are immediately identifiable by the severe inflammation and oedema at the site of infection. Examination of these regions will reveal a sloughing of the mucosa as well as large numbers of parasite stages.

The endogenous stages of *E. irresidua* are also found above and below the epithelial cell nucleus and in the submucosa. There are two schizogonic cycles according to Rutherford (1943) and the prepatent period is 9–10 days. *E. perforans* and *E. media* are two more small intestinal species, mild to moderate in the scale of pathogenicity. The endogenous stages of the former are always above the epithelial cell nuclei whereas *E. media* is found above and below the nucleus and in the submucosa. Of those species found in the caecum, *E. caecicola* is known to be non-pathogenic but *E. intestinalis* is very pathogenic. Details of species such as *E. piriformis*, *E. elongata* and *E. neoleporis* are not known.

Isospora sp.

The *Isospora* are also a cyst-forming group, having a cycle of asexual generations in epithelial cells and a very resistant oocyst. The principal difference from the *Eimeria* is that some *Isospora* have been shown to have tissue as well as gut stages (Frenkel & Dubey 1972) and these have been seen to invade the liver, spleen and lymph node. Structural differences can be seen in the mature cyst where the sporozites are neatly packaged into bundles of four within a distinctive sporocyst envelope. *Isospora* have been described in a range of mammalian hosts and are common in cats, dogs and man. A few have been reported in rodents, such as *I. ratti* in a rat in the USA (Levine & Ivens 1965), but only the sub-spherical oocyst measuring 22–24 × 20–21 μ has been described.

Disease in piglets associated with *I. suis* has been reported (Lindsay *et al.* 1980). Despite a broad scattering of reports from non-human primates, the pathology of infection in this group has not been well documented. Organisms have been described in a captive chimpanzee in Holland (Rijpstra 1967), in *Galago* sp. (Poelma 1966), Uakari monkey (Arcay-de-Peraza 1967), *Papio* sp. (McConnell *et al.* 1971) and capuchins and marmosets (Porter 1972; Hendricks 1974; Hsu & Melby 1974). A cycle of infection with common *Isospora* species may develop between cats and dogs excreting the cyst stage and susceptible rats and mice (Dubey 1978; Dubey & Mehlhorn 1978) (Fig. 3.13, page 69). This must be kept in mind where these animal species are in contact. The same goes for other cyst-forming species such as *Besnoitia wallacei* (Frenkel 1977) where schizogony and gametogony occur in the cat intestine and cysts are found in the fibroblasts of rats and mice. *Hammondia hammondi* has a cyst stage in the skeletal and cardiac muscle of mice, hamsters, guinea pigs and other rodents (Frenkel & Dubey 1975), but differs from *Toxoplasma*—where the parasite can be maintained within a single host indefinitely—by having an obligatory two-host cycle (Dubey & Wong 1978).

Toxoplasma sp. and Sarcocystis sp.

Two closely related members of the family Sarcocystidae, *Toxoplasma* and *Sarcocystis*, are of concern in laboratory animals (Levine 1961; Levine & Ivens 1965). A feature of all members is the mode of their deep tissue multiplication, known as endodyogeny, which is a process by which two daughter organisms are formed within a mother cell which is destroyed in the process. In the case of *T. gondii*, the merozoite production during the intracellular intestinal stages results in a double number of daughter cells, a condition known as endopolygeny.

The gametogenous stages of *T. gondii* in the cat take place in the epithelial cells of the ileum. The oocyst when released is very small (Fig. 3.14, page 69) measuring only 12 × 10 μ in comparison with *I. felis* (32–53 × 26–45 μ) or *I. rivolta* (20–27 × 15–24 μ), common coccidial parasites of the cat. When sporulated the oocyst contains two sporocysts each with four sporozoites. When ingested by a suitable host, which in this case is any mammal, the sporozoites hatch in the duodenum, invade the host cells and proceed to develop. Initially they are seen in the cells in pairs (endodyogeny), but once released from the cells they invade neighbouring cells and are dispersed around the body via the blood and lymph streams. Early localized infection is thus quickly generalized. Once the host immune responses are established the organism can be found as true cysts in areas of low immunogenicity, such as the central nervous system and the eye. In these protected situations they can remain for many years, even throughout the lifetime of the host. Problems then occur only when for some reason the immune defences of the host are lowered.

The major danger of infection with this organism is when a primary infection occurs in a pregnant animal, because during the phase of dispersion the organisms can cross the placental barrier and become established in the embryo. Depending on the stage of gestation, this may cause abortion or severe damage to the developing fetus. This situation has been very well documented in sheep (Beverley & Watson 1961, 1962; Watson & Beverley

1971; Dubey & Sharma 1980) and in man (Desmonts & Couvreur 1974; Desmonts *et al.* 1985). In the UK as many as 3000 women per year become infected for the first time during pregnancy, resulting in clinical problems for about 500 babies. This reaction to infection occurs once only and resulting immunity is solid.

In the animal house all the stock and the staff are at risk if cats in the same unit are excreting *Toxoplasma* oocysts. For the cat the prepatent period is approximately 4 days and patency lasts up to 20 days after infection with brain cysts (Dubey 1976). Millions of oocysts are shed during that time and as the infection is not self-limiting as in *Eimeria* species, a re-shedding can occur shortly after infection with another species such as *I. felis* (Fleck *et al.* 1972; Dubey 1976). *Toxoplasma* has been reported in a wide range of non-human primate species and resurgence of infection has been reported in macaques (Wong & Kozek 1974).

Propagation of *Sarcocystis* occurs in a muscle cyst located in the skeletal or heart muscle of the intermediate host (Fig. 3.15, page 70). These cysts are ellipsoidal and develop within the cell without affecting the cell nucleus. The morphological features of the cyst are distinctive and identification is based on the size of the cyst and the structure of the cyst wall. Within the cyst a mass of cigar-shaped spores or bradyzoites can be seen. These infected muscles, of the mouse for example (Powell & McCarley 1975), when fed to the cat result in the development of gametogenous stages in the small intestine (Marcus *et al.* 1974; Ruiz & Frenkel 1976). After a prepatent period of 5–13 days, oocysts are shed in the stool. These measure $31–42 \times 25–33\,\mu$ and when sporulated contain two sporocysts each with 4 sporozoites around which the thin wall of the cyst often contracts giving it a dumb-bell appearance. Sporulation takes place at around 2–3 days under suitable conditions such as a covered petri dish at room temperature. When fed to mice the sporozoites hatch in the duodenum and find their way to the target muscles and penetrate the muscle cell. The mature cyst measures $150\,\mu$ in diameter and extends the full length of the muscle fibre, although it is about three months before it is sufficiently developed to be seen by the unaided eye.

Ninety-three species of *Sarcocystis* have been recorded (Levine & Tadros 1980) from a range of rodents including rats, mice, hamsters and guinea pigs. *Sarcocystis* has also been well documented in primates (Lussier & Marois 1964; Mandour 1964, 1969). Karr & Wong (1975) found that of 375 wild-caught Old and New World monkeys, 79 (21%) were infected.

Cryptosporidium sp.

First described by Tyzzer (1910, 1912) in the stomach and small intestine of the mouse, cryptosporidia (family Cryptosporidiidae) are now known to be present in a range of mammals and birds, including primates, rats, guinea pigs and rabbits. The organism is a small body attached to the surface of epithelial cells lining the stomach and intestine, but has been reported in the biliary and pancreatic ducts as well as the intestines of rhesus monkeys (Kovatch & White 1972; Cockrell *et al.* 1974). All the stages are attached to the cells by means of a knob-like organ, providing a very close attachment which is best seen using transmission electron microscopy (Vetterling *et al.* 1971). The oocysts are smooth-walled, measure $7 \times 5\,\mu$ and do not contain sporocysts, residual body or micropyle. The schizonts, similarly attached, usually contain eight zoites, all of which are attached at one end, have a slight spiral twist and a globule of residual material near the point of attachment.

The gamonts are smaller at $5 \times 3.5\,\mu$. The microgametocyte contains a large residual mass and 16 microgametes which consist of a rod of chromatin with a small amount of faintly staining material trailing from them but no flagella. The macrogametocytes contain oval refractile granules which stain reddish-brown to deep purple with iodine and red with Giemsa after formalin fixation. After fertilization the wall of the macrogametes, unlike other stages, becomes thickened forming a resistant oocyst, a large proportion of which is filled with food granules. The protoplasm divides into four sporozoites while still within the host cell.

Eighteen species of *Cryptosporidium* have been named since Tyzzer's first description, but the likelihood is that they are not host-specific. Tzipori *et al.* (1983) were able to cross-transfer cryptosporidia from calves and suggested a single species gene. Levine (1984) reviewed the topic and suggested four valid species. Upton & Current (1985) have described a species causing diarrhoea in dairy and beef cattle which is morphologically indistinguishable from *C. parvum* (Tyzzer 1907, 1912). The organisms are also reported to cause enteritis in the guinea pig (Jervis *et al.* 1966) and the cat (Bennett *et al.* 1985) as well as man (Current *et al.* 1983; Upton & Current 1985). Recent interest in this parasite has revealed it to be a very widespread cause of enterocolitis. Isolation and identification are carried out from faeces using Sheather's sugar concentration method or smears stained by Giemsa (Tzipori *et al.* 1983), modified Ziehl-Neelsen (Henriksen & Pohlenz 1981; Garcia *et al.* 1983) or by safranin-methylene blue (Baxby *et al.* 1984) (Fig. 3.16, page 70).

Klossiella sp.

Members of the genus *Klossiella* have large ellipsoidal double-walled oocysts. Some of these contain numerous spherical sporocysts each containing 3–10 sporozoites. In some species, such as *K. muris* found in the mouse, it is difficult to ascertain whether there is an oocyst wall produced by the parasite or not (Wenyon 1926; Reichenow 1953; Levine & Ivens 1965).

Schizogony takes place mainly in the kidney glomeruli and endothelial cells of the kidney arterioles, although stages have been described in a wide range of other tissues (Twort & Twort 1923). There are possibly two types of schizont described by Stephenson (1915): the first contains 8–12 merozoites; the second, in the endothelial cells of the kidney capillaries and arterioles, contains 40–60 merozoites measuring $7 \times 2\,\mu$. The host becomes infected by ingestion of the sporocyst, the sporozoites then emerge, penetrate the mucosal wall and are disseminated by the bloodstream (Stephenson 1915). The schizonts occupy a large part of the Bowman's capsule which ruptures, releasing them into the renal space below the kidney capsule.

The merozoites enter the epithelial cells of the convoluted tubules of the kidney and become gamonts (von Sternberg 1929; Gard 1945). Micro- and macrogamonts occupy the same vacuole in the host cell. The microgametocyte forms two microgametes, one of which fertilizes the macrogamont and the resulting large zygote can be $40\,\mu$ in diameter. Within the zygote the sporont forms 12–16 sporoblasts each containing 25–34 slender sporozoites lined up in little bundles and a sporocyst residual body. The mature sporocyst is spherical, thin-walled and measures $13–16\,\mu$. It is released into the kidney tubules and is passed out in the urine. Nothing is known of the prepatent period but as the organism has only been found in older mice, it could be a long time. There have been no reports of infection in caesarean-derived mice and there have never been any reports of fatalities due to infection despite extensive damage to the kidneys.

A similar species, *K. cobayae*, has been described in the guinea pig. The first schizogony takes place in the endothelial cells of the capillaries, usually those of the kidney glomerulus. Each sporozoite penetrates a cell and rounds up to form a $2.5\,\mu$ trophozoite. Eight merozoites are formed from the divided nucleus of the trophozoite; they measure $2 \times 1\,\mu$ and cause the cell to bulge into the lumen of the capillary. They rupture into the vessel and are transported in the blood, invade other endothelial cells and repeat the cycle. Some merozoites reach the kidney tubules where they penetrate the cells and become schizonts which produce around 100 merozoites. Gametogony takes place in the loops of Henlé, the gamonts in pairs as before, sometimes with multiple microgametes. Sporulation occurs, the sporocysts having 30 sporozoites. This species has been reported only in the guinea pig and has never been known to cause clinical symptoms. Pathology of the kidney is slight, but Bonciu *et al.* (1957) claimed that 88% of cases in their laboratory went on to develop subacute nephritis.

Haemosporidia

Plasmodium sp.

Malaria parasites of the genus *Plasmodium* are likely to be found only in wild-caught monkeys and apes. Many species exist in rodents in the wild, but no instance of natural infection has been reported in laboratory rodents although they are commonly used as model hosts for the screening of drugs. The slender sporozoites are introduced into the bloodstream of the mammal host while the insect vector, a mosquito, is feeding. They circulate briefly in the peripheral blood and settle within the parenchymal cells of the liver. They develop into a meront over the next 1–2 weeks, known as the pre-erythrocytic schizont, containing thousands of merozoites and gradually filling the cell which eventually ruptures. The released merozoites enter the bloodstream where they commence the erythrocytic phase, entering the red blood cells and forming firstly a characteristic ring shape, then a trophozoite which undergoes nuclear division forming an erythrocytic schizont. This begins a new cycle of asexual multiplication as the cell bursts and the merozoites seek new red blood cells. The next stage is the sexual stage when some of the merozoites entering red cells become either male or female gamonts.

Distinguishing between the species of *Plasmodium* relies on the characteristic features of erythrocytic schizont cycle and stages. The sporozoite in the cell often causes a change in shape and produces distinctive stippling. The shape of the developing trophozoite is also specific for each species from the time of the ring form. The pattern of the disease depends on the rate of development culminating in rupture of the erythrocytic schizonts. This gives a characteristic fever pattern of 24, 48 or 72 hours duration due to parasite and cell debris being released into the circulation. In its primary host the infection comes under control with the development of host immune response and there are no fatalities. In some species no pre-erythrocytic forms remain in the liver; in others, some merozoites reinvade parenchymal cells and remain as a reservoir causing a relapse of infection months later. The most obvious symptoms are the recurrent fevers and the enlargement of liver and spleen. Histologically malarial pigment can be found in the reticuloendothelial cells of both these organs, which may also contain large numbers of mature schizonts as schizogony often takes place in the finer vessels of major organs such as the brain, heart and liver.

Plasmodium cynomolgi (Fig. 3.17, page 71) is common in the cynomolgus monkey and a range of other macaques and is only mildly pathogenic. The schizont cycle is 48 hours and the short thick ring form and trophozoite cause the red blood cell to enlarge. This is a characteristic feature, together with the coarse stippling in the cell. *P. knowlesi* is also a common parasite of the same hosts, leaf monkeys and man (Fig. 3.18, page 71). It has a 24-hour cycle and often multiple infections are found in the infected red blood cells which are finely stippled. This is a much more pathogenic species and a very virulent strain can be fatal to macaques. *P. inui* has a 72-hour cycle and produces very small rings which appear to have little effect on the red blood cell which is also finely stippled. This is known as quartan malaria with its three-day cycle of chills and fever but it is generally considered to be only mildly pathogenic. It has been observed to persist for several years in captive animals.

Various species of *Plasmodium* from wild rodents are used in malaria research. *P. berghei* from the tree rat is transmissible to rats and mice, *P. vinckei* from African rodents to the mouse, and *P. chabaudi* from the tree rat to the mouse, where it produces a fatal malaria. However, when the mouse is simultaneously infected with the malaria and the rickettsial parasite *Eperythrozoon coccoides*, only a mild chronic malaria is produced (Ott & Stauber 1967).

The sexual stages of *Plasmodium* sp. mature in the stomach of the mosquito vector, usually from a range of *Anopheles* species. Once in the stomach the microgametocytes mature and produce a number of whip-like microgametes by means of exflagellation. One of these fertilizes the macrogamete, which becomes an active zygote capable of migrating through the stomach wall to its outer surface where it grows into an oocyst filled with thousands of fine spindle-shaped sporozoites. When fully mature these emerge from the oocyst and migrate to the salivary glands. The mosquito is now infective and the sporozoites are injected into the next host as it feeds (Garnham 1966).

Hepatocystis sp.

Only the sexual forms of *Hepatocystis*, another malarial parasite, are found in the peripheral blood (Fig. 3.19, page 71). Asexual multiplication is limited to the exoerythrocytic stages found in the liver. The meront develops slowly to full maturity, taking two months to reach a size of 1–2 mm in diameter (Garnham & Pick 1952; Miller 1959; Garnham 1966). The centre of the cyst or megaloschizont is filled with fluid and around the convoluted cyst wall thousands of developing merozoites can be seen. Eventually the cyst bursts and the released merozoites enter the red blood cells where they develop into gamonts over a period of 4–5 days. The vector of *Hepatocystis* is *Culicoides*, a midge (Garnham *et al.* 1961). The cycle in the insect is similar to that found with *Plasmodium* in the mosquito, with the development of an oocyst in the haemocoele over a period of five days. The sporozoites emerge from the oocyst, migrate to the salivary glands and enter the next host at the time of the next blood meal. *H. kochi* is a common parasite of African green monkeys, macaques and baboons (Kuntz *et al.* 1967*a*, *b*). It is non-pathogenic, with no clinical signs, and the extensive liver stages are easily observed in histological sections.

Babesia sp.

The babesias are piroplasms, small and unpigmented intra-erythrocytic protozoa which are found in a wide range of mammals. They are severe pathogens of domestic animals and dogs, and in rodents the species *Babesia microti* is zoonotic, being isolated from splenectomized (known as the European form) and unsplenectomized humans (American form) (Gleason *et al.* 1970; Ruebush *et al.* 1977, 1979; Anderson *et al.* 1979; Cox 1980; Roth *et al.* 1981), other primates (Shortt & Blackie 1965: Ruebush *et al.* 1979; Moore & Kuntz 1981) and from splenectomized and congenitally asplenic mice (Irvin *et al.* 1981). In hamsters there is a high proportion of infected cells, resulting in their being the most commonly used model for biochemical and immunological studies of the disease.

Within the red blood cell, division takes place by binary fission and the parasites appear amoeboid or ring-shaped within the cell, affecting its shape—as do the plasmodia—by causing enlargement depending on the species. *B. decumani* and *B. rodhaini* are species isolated from wild Norway rats and tree rats respectively. Both are widely used in the study of malaria in laboratory rats and mice where the infection reaches a peak 6–13 days post-infection. *B. canis* is a common pathogen of dogs in the tropics where the brown dog tick, *Riphicephalus saguineus*, is the usual vector. *B. pitheci* is a non-pathogenic species of macaques, baboons and other monkeys in Africa.

Entopolypoides sp.
Entopolypoides macaci is pathogenic and has been isolated from cynomolgus and African green monkeys. In all species of the parasite, once it is ingested by its secondary host, the tick, development is very similar. The infecting cell becomes vermiform and enters the cells of the intestine where it undergoes a multiplicative stage forming over a thousand individuals in a period of 2–3 days. These migrate into the body cavity, find their way to the ovaries and enter the egg cells where they divide once or twice. They remain quiescent until the larval tick develops into a nymph, at which point they migrate to the salivary glands where they multiply once again, filling the cells with tiny parasites which emerge into the lumen of the salivary ducts and are passaged into the next host when the tick feeds. No sexual stages have been observed.

Microsporidia

Encephalitozoon (Nosema) cuniculi is a neosporan with an ovoid or rod-shaped body and an amoeboid sporoplasm found in cysts or free in the body tissue. It is the only species of this genus known in endothermal animals where it has been reported in a wide range of host species including man (Levine 1961; Barker 1974; Gannon 1979; Wilson 1979). In the laboratory its most important host is the rabbit, in which it produces extensive granulomas in the brain and damage to the kidneys (Fig. 3.20, page 72).

The trophozoite measures 2–6 μ in length and 1–3 μ in width with an eccentric nucleus filling a quarter of the body. The non-dividing spore form measures $2.5 \times 1.5\ \mu$, has a firm capsule which is strongly Gram-positive and a neatly coiled filament and polar cap. Transmission usually occurs by oral contamination of a new host with spores from the urine of another.

Once in the alimentary tract the spore comes into close contact with the wall of the mucosa and the intricate firing mechanism comes into action, the sporoplasm being shot into the cytoplasm of the host cell (Lainson *et al.* 1964). Multiplication is by means of a form of schizogony and the cells of the reticuloendothelial system which engulf the intruders are themselves invaded. In the initial proliferation of the infection the trophozoites can be seen widespread throughout the body within macrophages and nerve cells where there can be up to 100 organisms in what are known as pseudocysts.

Mice artificially infected with the organism show an increase in numbers of organisms for up to two weeks when the infection reaches a peak, with ascites and abdominal enlargement around two weeks post-infection. Thereafter organisms become localized in the brain and kidney. In rodents these cysts can be 20–45 μ in diameter. During the early part of the infection in rabbits, large numbers of spores may be found in the urine. In the brain, extensive granulomas up to 1 mm in diameter can be found (Goodpasture 1924) but few spores can be seen at this stage and then normally only with Gram stain (Goodman & Garner 1972). In the chronic state the kidney is usually enlarged with multiple scarring and pits (Koller 1969) and may be pale in colour. The spores are found in the tubular epithelial cells and when mature are flushed out with the urine (Yost 1958).

There has been considerable speculation on the possibility of transplacental transmission, and this may well have been the cause of infection in caesarean-derived rabbits as described by Innes *et al.* (1962). However, it has not been observed by the author even in rabbits given infective spores just before mating or at mid and late pregnancy (Owen & Gannon 1980; Wilson 1986). In an extensive survey by Gannon (1979) it was found that all the large commercial rabbit colonies at that time were infected by this organism. He also observed that in animal houses without rabbits the rodent colonies were not naturally infected although all were susceptible to infection. The level of infection within the rabbit colony is directly related to the level of hygiene, and the main route of infection between the rabbit and other species such as rats, mice, hamsters and guinea pigs is via contact with the urine of infected rabbits. Young rabbits from infected mothers, if fostered from birth onto uninfected mothers, will remain free of infection. Even weanlings from infected mothers, if they are removed to a clean environment, can remain infection-free (personal observation). In a group of infected mothers and offspring, a rising antibody titre was observed at two months of age in those animals remaining in the infected animal room.

Despite being a common parasite of the rabbit, *E. cuniculi* rarely causes any clinical disease even when there are quite extensive brain and kidney lesions. In young rabbits, however, cases of slow growth and a roughened coat have been reported (Robinson 1954). The infection has been thought by some to be responsible for nervous disorders and lack of coordination as well as hind leg paralysis (Wright & Craighead 1922; Pattison *et al.* 1971). Histologically lesions can be seen in all the species vulnerable to infection. In the mouse examined at two weeks following an experimental infection, organisms were seen in the heart, lung and even in sections of the serosal coat of the small intestine (personal observation). These multiplicative stages rapidly disappeared; thereafter, cysts became apparent in the brain and there was a diffuse meningoencephalitis and perivascular infiltration. Only in the rabbit are there massive brain granulomas, kidney damage and an enormous number of excreted spores in the urine.

There have been only a few reports of microsporidia in primates. A *Nosema* sp. was reported in *Callicebus moloch* (Siebold & Fussel 1973) and also in the squirrel monkey, *Saimiri sciureus* (Brown *et al.* 1973), but it is not certain to date that this is the same species that infects rabbits.

ORDER RICKETTSIALES

The Rickettsiales are an odd group of organisms of uncertain classification and they are more closely linked to bacteria than to the Protozoa. However, they are important in the present context as potential disease-producing organisms, being transmitted between their rodent hosts by blood-sucking arthropods.

Haemobartonella and the closely related blood parasite *Eperythrozoon* have been classified as belonging to the Rickettsiales (Kreier & Ristic 1968, 1974). Both species are coccoid-like bodies measuring 0.3–0.7 μ (Small & Ristic 1967) and are transmitted by means of blood-sucking arthropods.

H. muris (Fig. 3.21, page 72) was once a common parasite of laboratory rats and could be seen closely applied to the surface of the red blood cell (Maede 1975; Maede & Hata 1975; Maede & Sonada 1975; Owen 1982) arranged in single or branched rows. The parasite causes the membrane of the red blood cell to invaginate so that the organisms are enclosed in channels on the surface; this accounts for the fact that fewer organisms are found free in a blood smear compared to *Eperythrozoon coccoides* of the mouse where the red cell wall does not invaginate. Both rickettsias are transmitted by species of blood-sucking louse, *Polyplax*, and are generally regarded as non-pathogenic, remaining as latent infections as long as the rat remains healthy. However, they appear in large numbers following splenectomy or concurrent infection with another blood parasite. Parasitaemia results in a severe anaemia which may prove fatal depending on the susceptibility of the host. *Haemobartonella* also causes a fatal anaemia in cats, and the fine structure has been described in the squirrel monkey (Aikawa 1972).

E. coccoides is a small disk- or ring-shaped organism measuring 0.4–0.6 μ in diameter which adheres closely to the surface of the red blood cells (Augsten 1982) in random clusters, and in blood smears can clearly be seen free between the cells (Fig. 3.22, page 72).

Following splenectomy both organisms appear on the red cells 3–7 days later and persist for several weeks. Anaemia is generally slight in natural infections of the intact animal, the only clinical sign being an enlargement of the spleen. In latent infection, *E. coccoides* is an effective activating agent of mouse hepatitis virus (Gledhill 1956) and lymphocytic choriomeningitis (LCM) (Seamer *et al.* 1961), and a suppressor of *Plasmodium berghei* (Peters 1965; Baker *et al.* 1971).

It has generally been thought that control of the ectoparasitic vector would be sufficient to control infection and it certainly has disappeared in modern caesarean-derived colonies. However, reports have been made of persistent infection, demonstrated by means of splenectomy, through several generations of mouse after the vector had been eliminated.

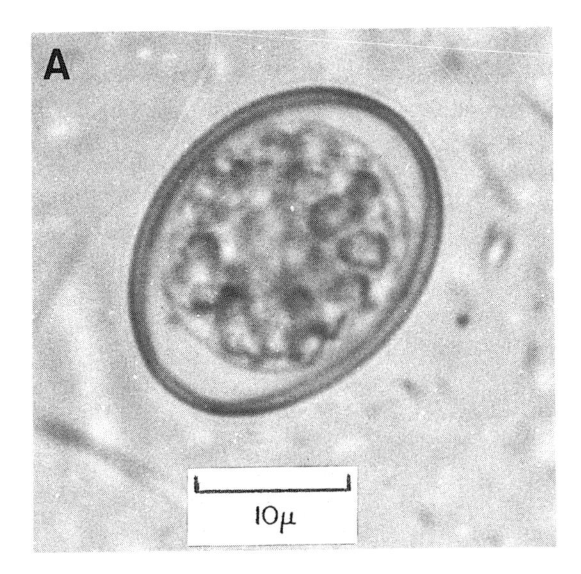

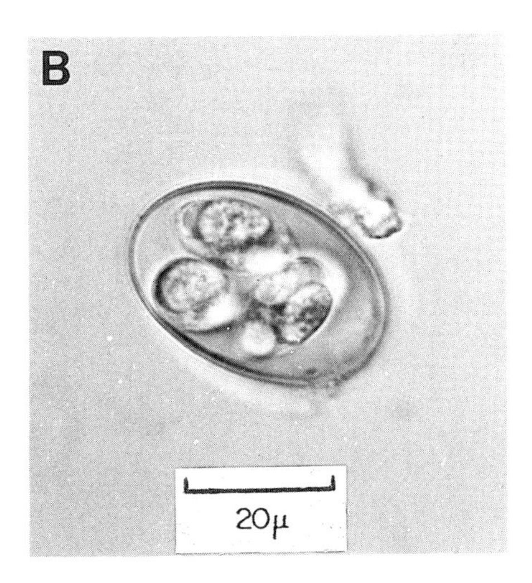

Fig. 3.9 *Eimeria* oocysts. **A**: Unsporulated *E. falciformis* from the mouse. **B**: *E. irresidua* from the rabbit. **C**: *E. caviae* from the guinea pig.

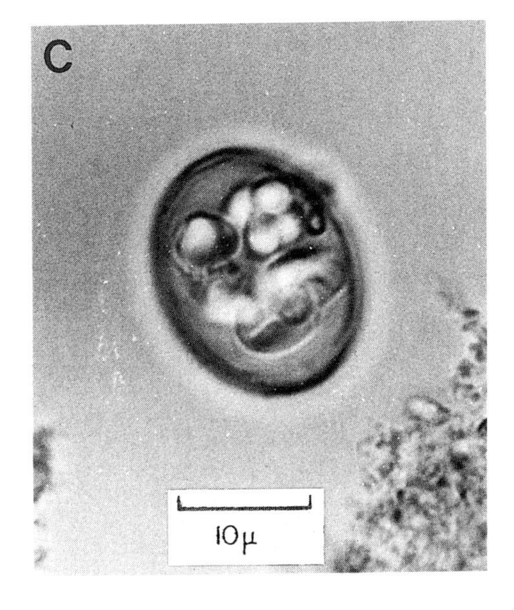

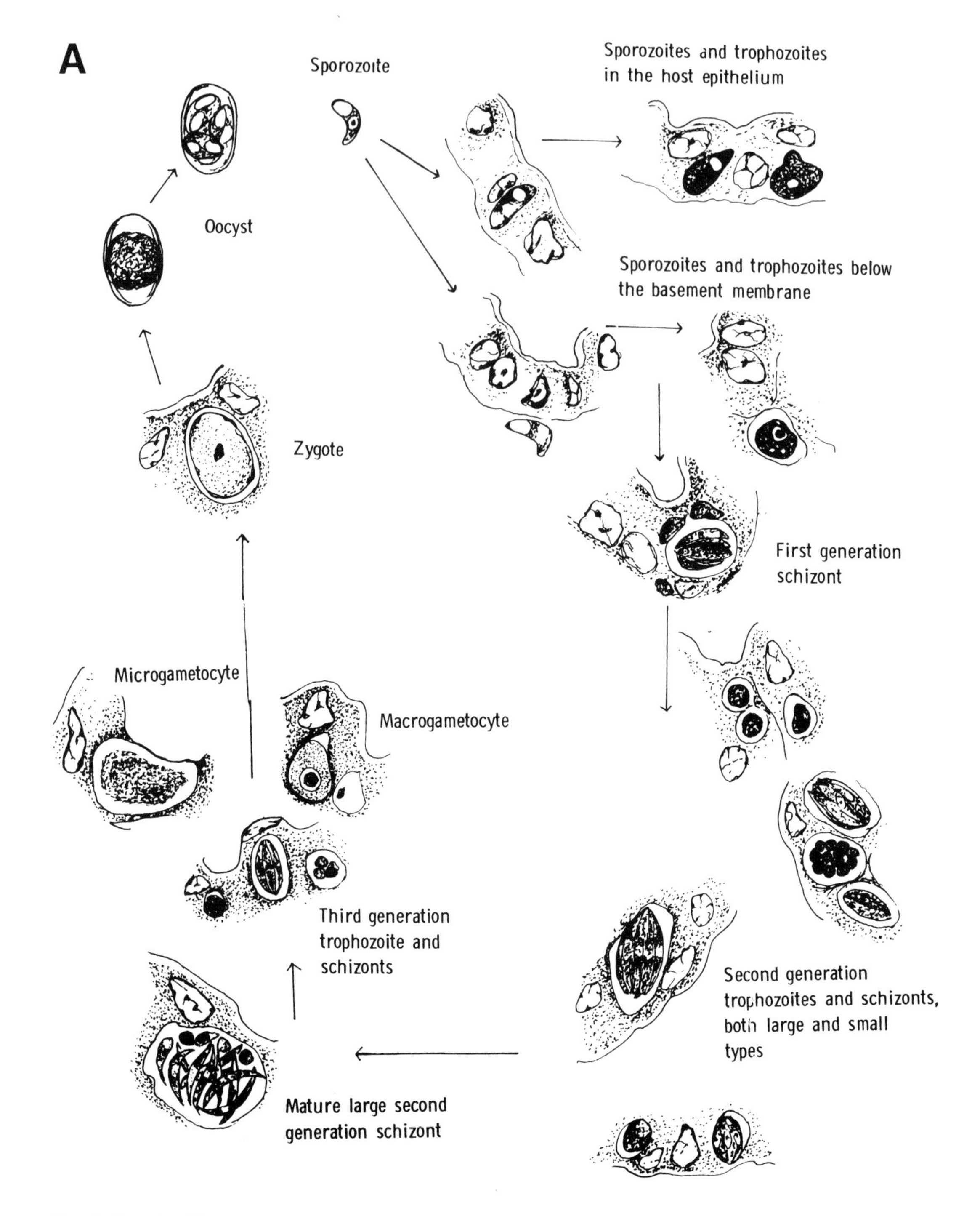

Fig. 3.10 The life cycle of *Eimeria falciformis* in the mouse. **A**: Diagram of the cycle. **B** (facing page): Histological demonstration of development.

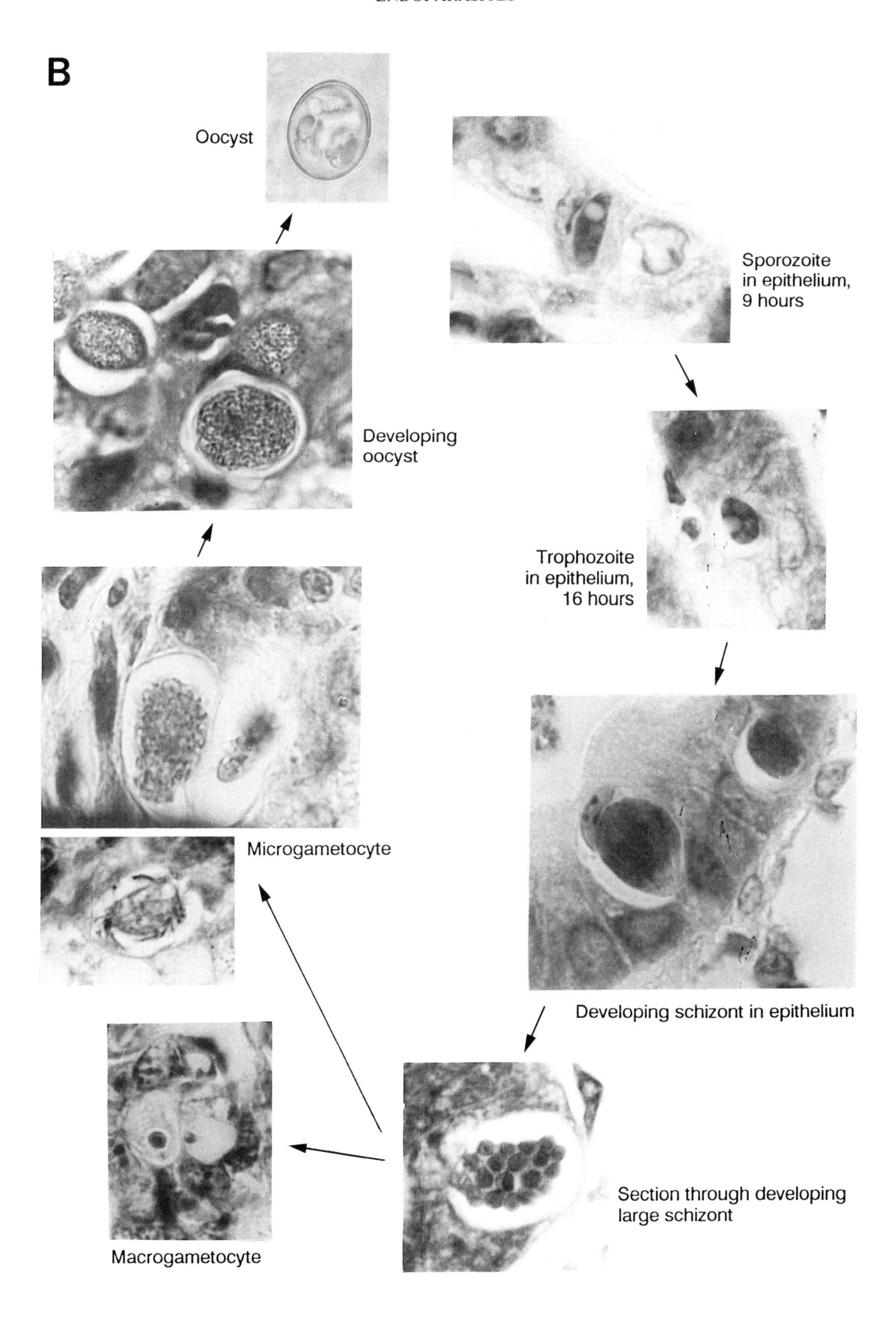
B
Oocyst
Sporozoite in epithelium, 9 hours
Developing oocyst
Trophozoite in epithelium, 16 hours
Microgametocyte
Developing schizont in epithelium
Section through developing large schizont
Macrogametocyte

Fig. 3.11 Intestinal *Eimeria* species of the rabbit, their prevalence and pathogenicity. (Adapted from Catchpole & Norton 1979, with kind permission).

SPECIES	OOCYST	PREDILECTION SITE	PREVALENCE	PATHOGENICITY
Eimeria magna	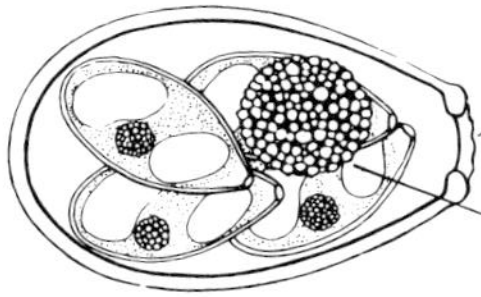 prominent micropyle with collar; large oocyst residuum	Jejunum and ileum	+ + + +	+
E. intestinalis	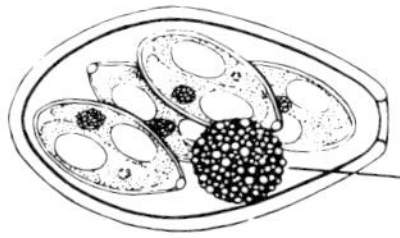 pyriform shape; large oocyst residuum	Jejunum and ileum	+	+ + +
E. piriformis	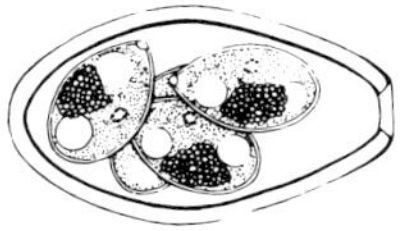 pyriform shape; NO oocyst residuum	Caecum and colon	+	+ + + +
E. irresidua	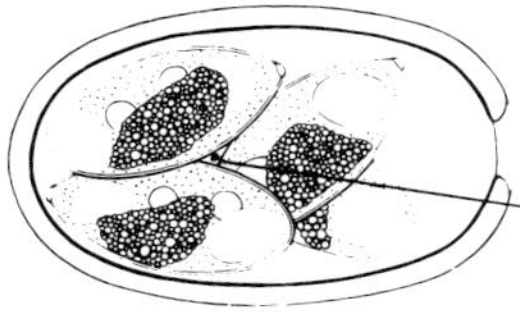 broad ellipsoidal shape; minute oocyst residuum usually hidden by sporocysts	Ileum	+ +	+

E. flavescens	ovoidal shape wide micropyle at broad end NO oocyst residuum	Ileum – 1st schizonts only Caecum and colon	+ +	++++
E. perforans	small colourless wall no micropyle 40 μm	Jejunum	+ + + +	+
E. media	micropyle a shallow pyramid large oocyst residuum	Duodenum and jejunum	+ + + +	+ +
E. coecicola	elongate ellipsoidal shape narrow micropyle large oocyst residuum	Ileum – schizogony Caecum – gametogony	+++	+
E. stiedae	elongate ellipsoidal shape narrow micropyle minute oocyst residuum usually hidden by sporocysts	Liver	+	+ + + +

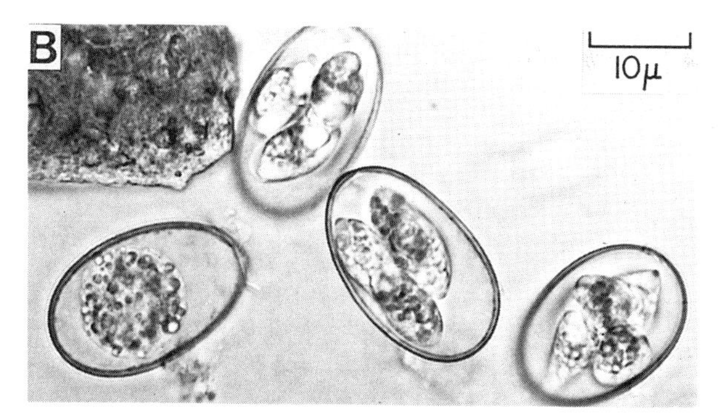

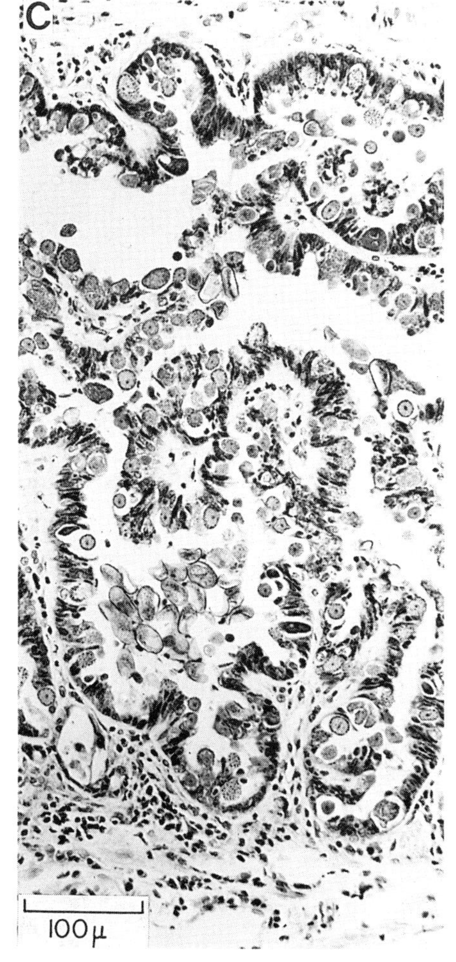

Fig. 3.12 *Eimeria stiedae*, the cause of liver coccidiosis in the rabbit. **A**: Severely affected liver with grossly enlarged bile ducts. **B**: Sporulated and unsporulated oocysts. **C**: Section through infected liver showing proliferation of the bile ducts containing schizonts and gametogenous stages.

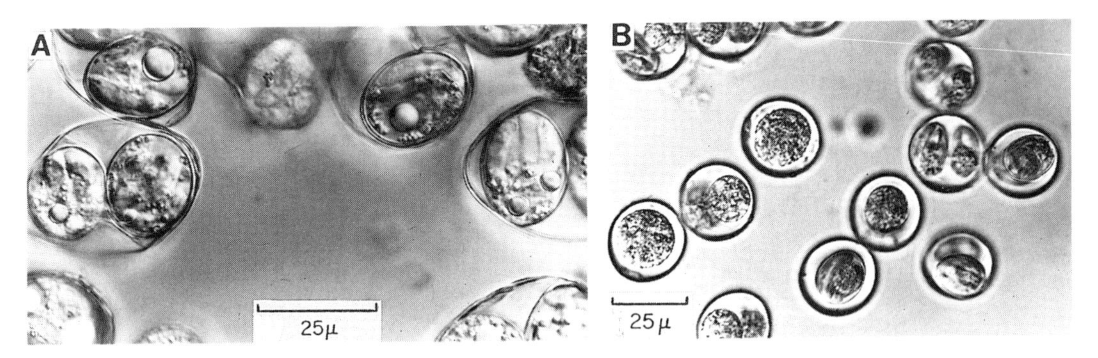

Fig. 3.13 *Isospora* species of the cat. **A**: *I. felis*. **B**: *I. rivolta*.

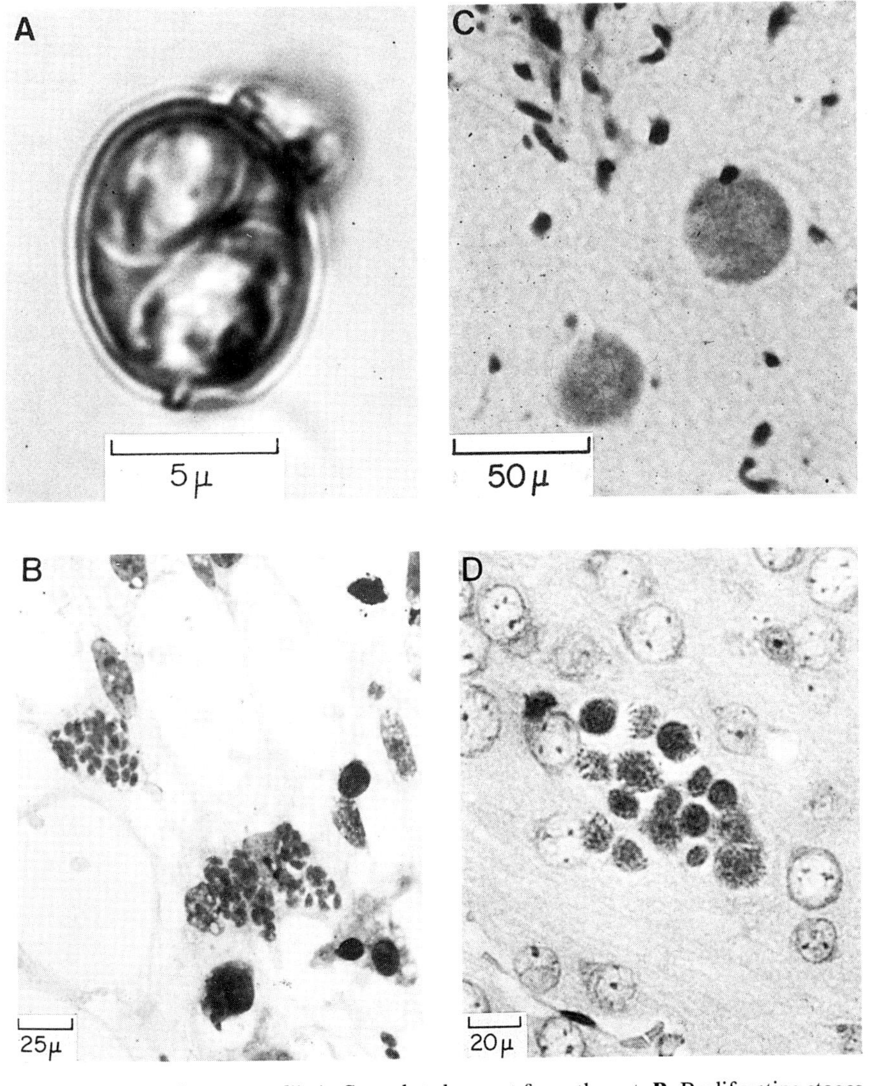

Fig. 3.14 *Toxoplasma gondii*. **A**: Sporulated oocyst from the cat. **B**: Proliferating stages in the mesentery of the mouse. **C&D**: Proliferating and cyst stages in the brain of a mouse.

Fig. 3.15 *Sarcocystis* cysts like pale threads in the muscles of a mouse.

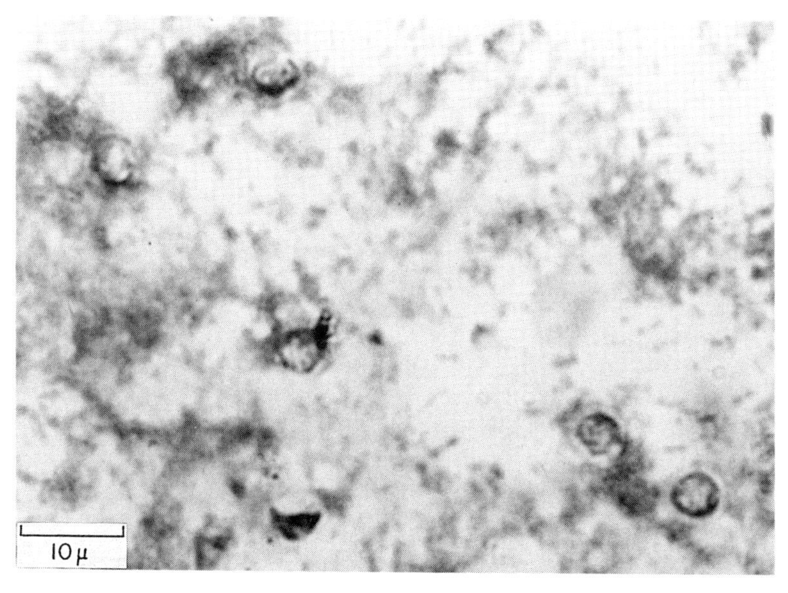

Fig. 3.16 Oocysts of *Cryptosporidium* sp. in faeces, demonstrating ease of observation with safranin stain.

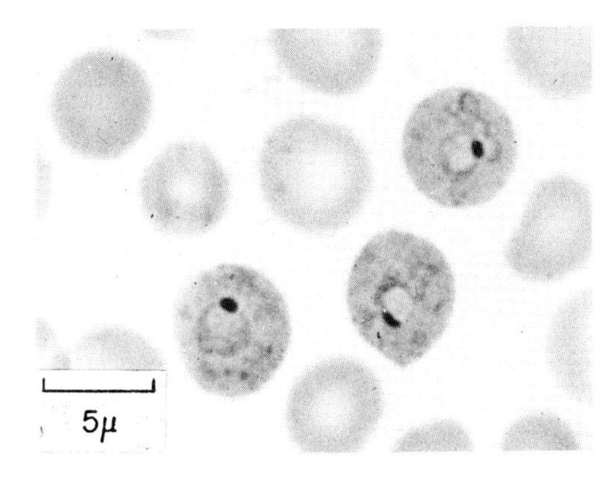

Fig. 3.17 *Plasmodium cynomolgi* ring forms in the blood of a macaque.

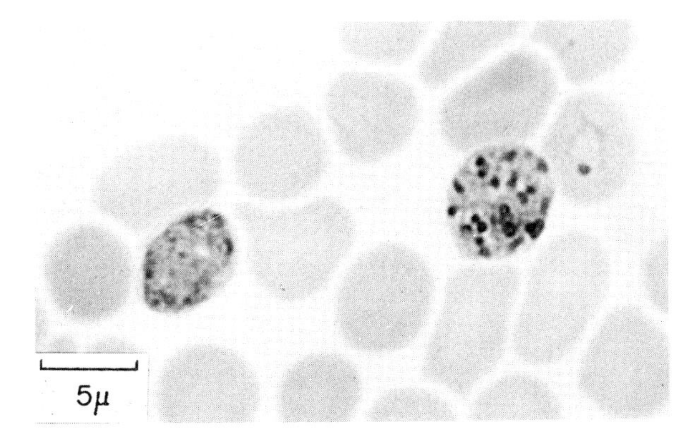

Fig. 3.18 *Plasmodium knowlesi* ring forms, trophozoite and schizont, in the blood of a macaque.

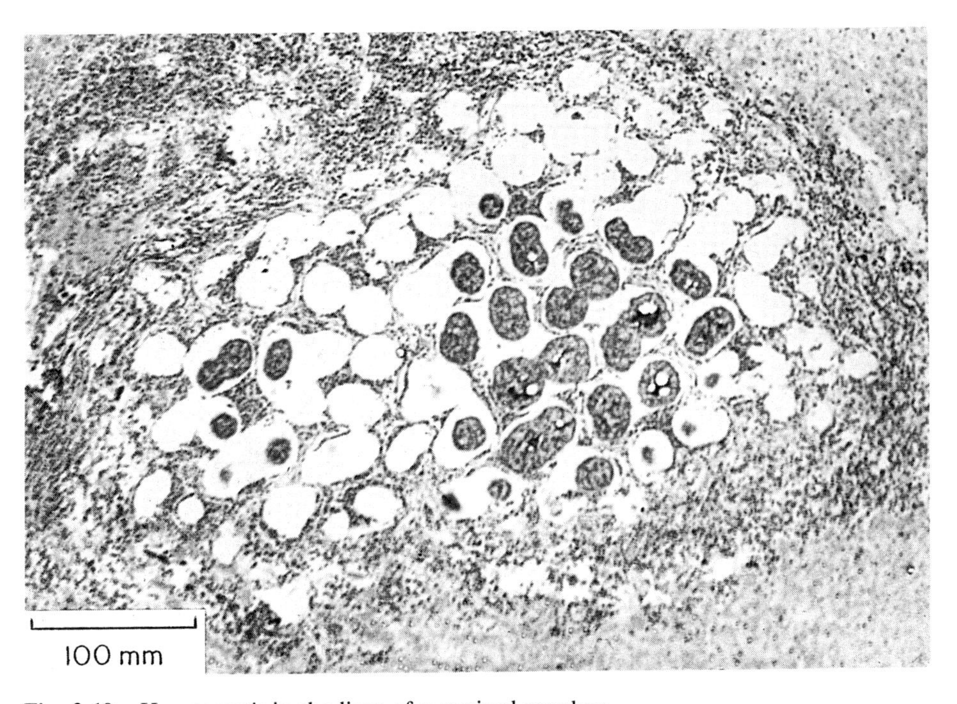

Fig. 3.19 *Hepatocystis* in the liver of a squirrel monkey.

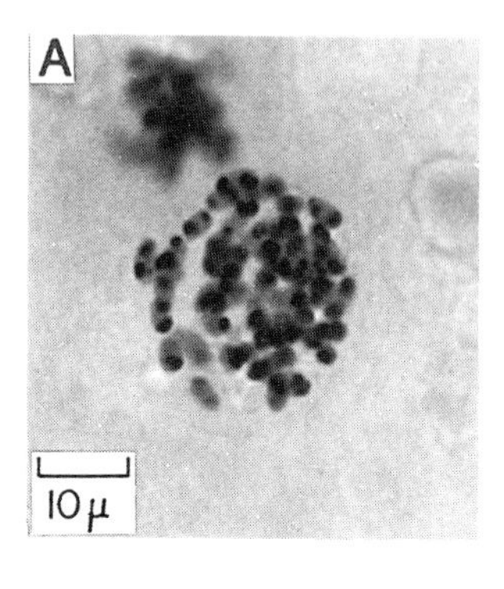

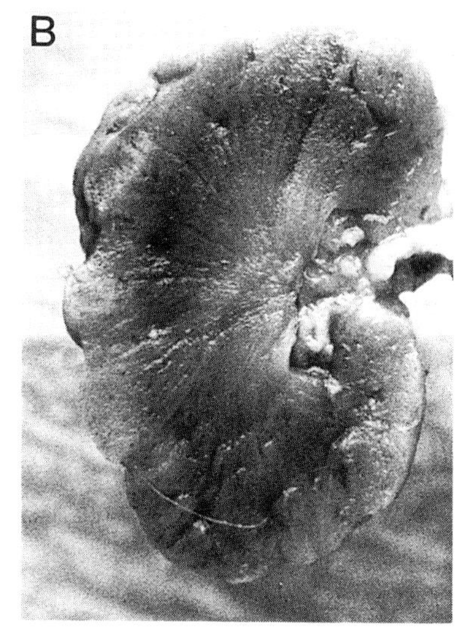

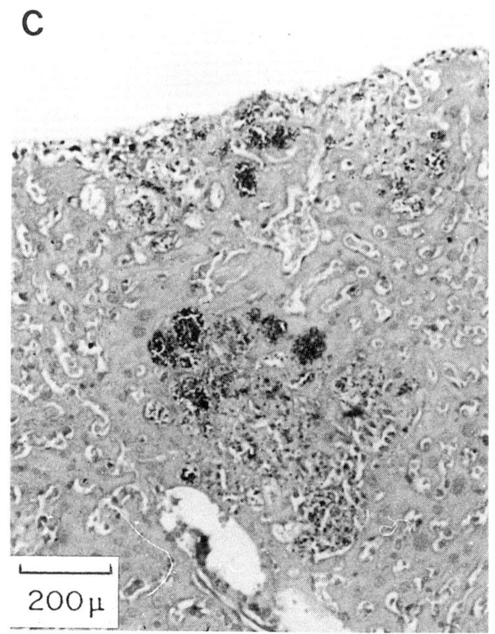

Fig. 3.20 *Encephalitozoon* (*Nosema*) *cuniculi*. **A**: Cell containing proliferating spores in the peritoneal cavity of a mouse during passage. **B&C**: Typical kidney lesions from an infected guinea pig.

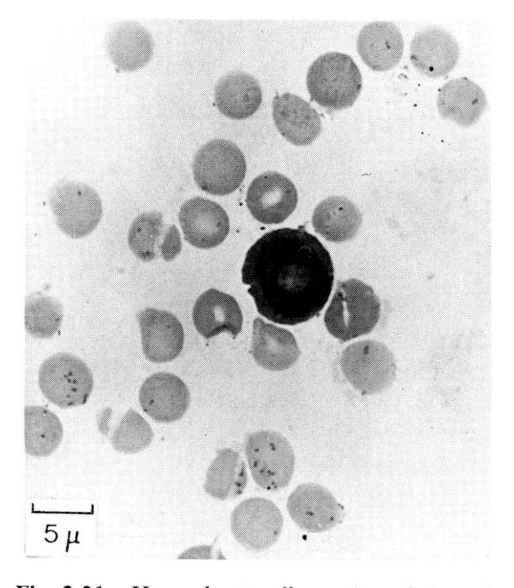

Fig. 3.21 *Haemobartonella muris*, rod-shaped bodies on the red blood cells of a rat.

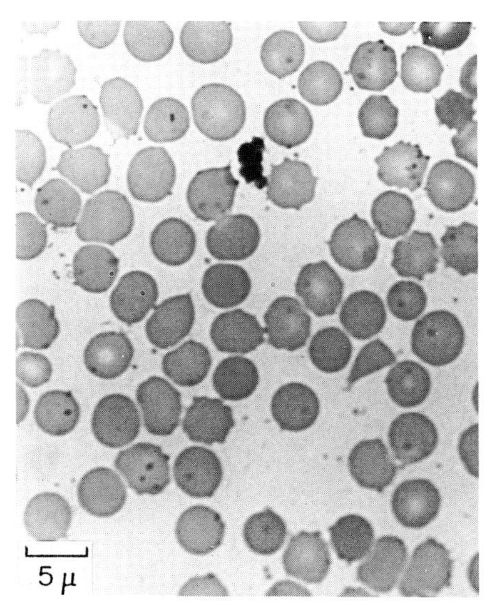

Fig. 3.22 *Eperythrozoon coccoides*, ring-shaped bodies on and between the red blood cells of a mouse.

PHYLUM PLATYHELMINTHES

CLASS CESTODA (Tapeworms)

Order Cyclophyllidea

The most common cestode parasites of laboratory animals are all members of the order Cyclophyllidea. Because of the complexity of their life cycle and the need for a secondary host, most cestodes are rapidly eliminated from the animal room when the standards of hygiene improve and the secondary hosts, usually arthropods, are excluded.

Family Hymenolepidae
The usual cestode inhabitants of laboratory rodents are members of the genus *Hymenolepis*. Of the three species reported from rodents, *H. nana*, *H. diminuta* and *H. microstoma*, the most persistent is *H. nana* because of the greater flexibility of its life cycle (Fig. 3.23, page 81). It is known as the dwarf tapeworm because of its small size, usually around 20–40 mm, although specimens of 10 cm in length by 1 mm wide have been reported. The size of tapeworms is strongly influenced by the number of individuals competing for space, together with the nutritional status of the host.

Morphologically *H. nana* is typical of its class, having a well-defined scolex bearing four strong suckers between which is a protrusable rostellum crowned, in this organism, with a circlet of small hooks. At the base of the scolex is a short neck from which develops the segmented strobila. The segments are wide and shallow and each is provided with both male and female organs. A mature segment contains three spherical testes, a single ovum and an extensive branched vitelline gland. Each segment has a unilateral genital opening and the progress of sexual maturity can be seen along the length of the worm from the immature first segments at the base of the neck to the terminal segments which are either packed with eggs or shrivelled and empty where all the eggs have been shed.

The egg measures 40–45 μ in length by 34–37 μ in width and has a thin-walled, clear outer shell. In the centre is the spherical embryophore, also thin-walled, with a distinct knob at each pole from which five or six fine trailers emerge giving it a distinctive appearance. The spherical onchosphere within the embryophore contains six distinctive embryonic hooks which can be seen clearly through the shell.

Hymenolepis adults are found in the lumen of the small intestine of the mammalian host, the scolex being embedded in the mucosa and the strobila hanging free in the intestinal lumen. The eggs are shed into the lumen of the intestine and are excreted with the faecal material. The next stage of development depends upon the infestation of the diet with a food pest such as the 'confused flour beetle', *Tribolium confusum*. The beetle ingests the egg and, once this is in the stomach, the onchosphere emerges and, using the six embryonic hooks, energetically tears its way through the stomach wall and into the body cavity, or haemocoele, where it rounds up to form a cysticercoid. This is a solid pea-shaped body in which the scolex is neatly folded. Further development requires the ingestion of the infected beetle by the definitive host. When it reaches the small intestine of the host the digestive enzymes stimulate the scolex to emerge. It then attaches to the mucosal wall, the neck elongates and the strobila begins to develop. The cycle is completed with sexual maturity and the production of eggs. This period of development is known as the prepatent period and varies from 14 to 24 days.

The remarkable success of *H. nana*, even in a closed and barrier-maintained rodent colony, is due to its ability to bypass the secondary host and infect its definitive host directly (Heyneman 1953, 1961). Unlike any other cestode species, the egg will hatch in the small intestine of the mammalian host and the onchosphere will burrow into a villus of the mucosa and proceed to develop into a typical cysticercoid (Hunninen 1935*a*, *b*, 1936). After 4–6 days the larva emerges into the lumen of the gut, the scolex attaches to the wall of the mucosa and development proceeds as before.

Low levels of infection appear to have little effect on the host, but inevitably the high population density found in animal rooms leads to heavy infestation, most commonly in young animals where it is manifested in poor development and slow growth. Young hamsters are often affected, and a pot-bellied syndrome found among weanlings is usually associated with a high population of *H. nana* as well as large numbers of protozoa such as *Hexamita* sp. and *Giardia* sp. Where infected animals are stressed, such as happens when they are transported long distances, the sheer numbers of emerging larvae can result in deaths (personal observation). *H. nana* is a fairly common parasite of conventional laboratory rats, mice and hamsters and a similar strain has been described in man, but there have been no reports of cross-contamination between man and rodents. The human strain of *H. nana* has, however, been described in non-human primates.

H. diminuta is another common species found in rats, mice and hamsters. Unlike *H. nana*, the cysticercoid cannot develop in the villi of the definitive host. Some 30 species of insect have been found to be intermediate hosts to *H. diminuta*, the most common being the flour beetles *Tribolium confusum* and *Tenebrio molitor* (Vogue & Hyneman 1957) in the laboratory, and also the rat flea *Nosopsyllus fasciatus* in wild rodents. *H. diminuta* is usually larger than *H. nana*, measuring 20–60 cm in length by 0.5–4.0 mm in width. The scolex is similar, with strong suckers and rostellum but no hooks. These tapeworms are easily found in the upper portion of the small intestine and the eggs are distinctive, being semi-spherical, 60–66 μ in diameter, with a dark outer capsule (Fig. 3.24, page 82). The typical embryophore within it is thin-walled, as with *H. nana*, but lacks the polar knobs and filaments. Transmission is always by ingestion of the cysticercoid when diet containing infected beetles is eaten. The prepatent period of this species is 19–20 days.

H. microstoma is a species found in wild mice but which has never been reported in laboratory species (Joyeaux & Kobozieff 1928; Dvorak *et al.* 1961; Litchford 1963). Morphologically it is very similar to *H. nana*, including the row of 23–28 hooks encircling the rostellum. The eggs are similar but slightly larger, measuring 80–90 μ with slightly fewer polar filaments. The adults measure 8–50 cm in length, but the most important difference between the two species is their location which in *H. microstoma* is deep in the bile duct (Fig. 3.25, page 82). The onchosphere, when it emerges in the upper small intestine, migrates up the bile duct and often into the finer branches. Thus the gall bladder may become filled with worms and in a heavy infection the bile duct and duodenum may become grossly enlarged. This dilation is thought to be due to a toxin produced by the worms and not just to crowding. Heavy infestations can result in considerable damage with necrotic lesions appearing on the liver surface and biliary cirrhosis may occur. Blockage of pancreatic ducts may also result in the formation of retention cysts in the pancreas. This species has not been reported in wild rats and experimental infections have shown that the rat is generally a refractory host, possibly because it lacks a gall bladder and loses its parasites early. The hamster, on the other hand, is extremely sensitive and pathological changes may prove fatal (Dvorak *et al.* 1961; Litchford 1963).

Family Taenidae

Another extensive family in this order is the Taenidae. The adults of this group are large worms inhabiting the intestinal tract (Lapage 1968), and the stages of the life cycle affecting laboratory rodents are the larvae in fairly large, fluid-filled cysts, which account for their being more commonly known as the bladderworms.

The larval stage of the cat tapeworm *Taenia taeniaeformis* was once common in conventionally bred rats, mice and hamsters. The adult tapeworm is an intestinal parasite of the cat and related carnivores. It measures 15–16 cm with a well-marked scolex and segments longer than broad. The eggs found in the faeces are small and spherical measuring 31–37 μ and have a dark-coloured striated embryophore containing a typical 6-hooked onchosphere (Burrows 1965).

The eggs are ingested by the rodent intermediate host, and when the larva hatches in the small intestine it makes its way to the liver with the aid of the embryonic hooks where it develops into a cysticercus larva, *Cysticercus fasciolaris*, over a period of 30 days. It is an unusual larva having an atypical development compared to other bladderworms. The scolex is not invaginated but has the typical strong rostellum armed with 26–52 hooks and four prominent suckers. The neck is not developed but the body behind the scolex is apparently segmented or pseudo-segmented, giving it the appearance of an adult worm were it not for the presence of a small terminal bladder. The larva, when relaxed, measures 6–12 cm, and at post-mortem examination can be seen macroscopically as a creamy white body lying coiled in the cyst beneath the capsule of the liver (Fig. 3.26, page 83). Usually only one or two parasites are to be found and resistance to reinfection is solid. However, multiple infection by 10–20 of these bladderworms has been observed in some mice and hamsters (personal observation). When the larva in the rodent host is ingested by the definitive host, the predator, both the bladder and the strobila are digested away.

Despite the large size of some of the cysts and the presence of more than one, the effect upon the rodent host is usually inapparent (Oldham 1967). However, an association between infection and the development of liver tumours has been pointed out by Bullock & Curtis (1924) and Jones (1967).

A more typical larval bladder cyst is found in the mesentery of the rabbit. This is *Cysticercus pisiformis*, the larval stage of *Taenia pisiformis* (Fig. 3.27, page 84), a small intestinal tapeworm of the dog and other wild carnivores. As with *T. taeniaeformis*, the scolex is strong and well-defined, this time with a double row of strong hooks on the rostellum. The mature segments are longer than wide and terminally these are packed with eggs which are shed into the lumen of the gut and passed out in the faeces. The eggs are typically taenid, being dark in colour, and spherical to sub-spherical, measuring 34–41 × 29–35 μ. In the small intestine of the rabbit host the onchosphere emerges from the egg and migrates first to the liver where it remains for three or four weeks. Next it migrates to the surface of the liver and then to the peritoneal cavity where the typical cysticercus larvae are easily located, being transparent rugby ball-shaped bladders with a distinct opaque body at one end which is where the inverted scolex is located. On ingestion by a suitable carnivore, the scolex evaginates in the small intestine and attaches to the gut wall where it develops into a mature worm within a few weeks. This was a fairly common parasite of rabbits when they were bred out of doors and at risk from contamination from dog faeces. Clinically the effects of light infection are inapparent, but heavy infections have been reported as causing abdominal distension and discomfort (Lapage 1968). Migration of the larvae through the liver results in the development of fibrous tracts and small necrotic foci can be seen. Only in massive infections is the liver likely to suffer severe damage.

Other bladderworm stages are more bizarre in their appearance. The adult stage of *Multiceps serialis* is an intestinal parasite of dogs and the larval stages are found in a range of mammalian hosts, usually wild rabbits and hares, but they have been reported in sub-human primates and man. The adult, *Taenia multiceps*, is a large worm measuring 70 cm by 3–5 mm (Fig. 3.28, page 84). The scolex has a muscular rostellum with a double row of strong hooks. The eggs are similar to those of other *Taenia* sp., measuring 31–34 μ (Burrows 1965). The onchosphere emerges in the small intestine and makes its way to the subcutaneous tissues. The fully mature larva is known as a coenurus type, measuring 4–5 cm in diameter; it is filled with fluid and has secondary buds protruding to the inside, each with an inverted scolex. The life cycle is very similar to that of *T. pisiformis* except for the multiple scolices and its position in the subcutaneous tissues. Pathologically this produces little effect in the rabbit, but fully developed cysts can be palpated as swellings under the skin (Cook 1969).

The most pathogenic larval species is *Echinococcus granulosus*, the small tapeworm of the dog and other wild carnivores and whose larval stage has been reported in most mammalian species. The tiny adults usually have only 3–4 proglottids and measure 2–9 mm. The scolex is typical of the Taenidae, with a strong scolex and rostellum topped with a ring of hooks. The eggs are also typical, ovoid in shape and measuring 32–36 μ. The emerging onchosphere migrates through the intestinal wall, enters the mesenteric blood vessels and is transported to the lungs or liver of the intermediate host where it develops into an enormous cyst 5–10 cm or more in diameter and filled with fluid. The unique nature of this cyst is its ability to produce secondary buds with inverted scolices. Unlike *Multiceps*, these multiple heads continue to increase in number together with invaginations of the cyst wall, producing daughter cysts each of which may contain more daughter cysts, all of which eventually sediment onto the bottom of the mother cyst.

The parent cysts vary in size and shape depending on the organ where they establish, and become a problem to the host when they have developed sufficiently to place pressure on other organs. The cyst can produce abdominal distension and distress in monkeys. Surgical removal of the cyst would be dangerous to the host should the cyst rupture, because of the ability of the released daughter cysts to continue developing. The effect of the adult parasites on the dog host is negligible despite the fact that they may be present in very large numbers.

These are not parasites likely to develop under modern laboratory breeding conditions. As with all these complicated cycles, they appear only when wild-caught animals or contaminated offal or fodder are brought into the building. Occasionally odd larval stages are found subcutaneously at post-mortem examination (Fig. 3.29, page 84) but they are rarely the cause of a clinical condition.

Family Cataenotaenidae

The only member of the Cataenotaenidae to be found in laboratory animals is *Cataenotaenia pusilla*, a small intestinal species common in wild rats and mice (Balfour 1922; Elton *et al.* 1931) (Fig. 3.30, page 85). It is a fairly large worm measuring 30–160 mm in length, its segments longer than broad with irregularly alternating genital apertures. The scolex has no rostellum and only weak, insignificant suckers. The terminal proglottids are packed with small, untidy, oval eggs measuring 22 × 12 μ (Joyeaux & Baer 1945) with the distinctive onchosphere within. While gravid proglottids separate into the intestinal lumen and can be seen actively moving about in the faeces, thus the individual eggs are rarely isolated in standard faecal flotation techniques. The intermediate hosts are the common grain mites *Glycophagus domesticus* and *Tyroglyphus farinae*; adults or nymphs become infected when rodent faeces containing eggs contaminate the diet. The onchosphere develops into a pleurocercoid of an incomplete type called a merocercoid (Joyeaux & Baer 1945). There is a single invaginated sucker at the apex and no lateral suckers as are found in other families of this order. The merocercoid is infective at 15 days post-infection, and once within the intestine of the definitive host the apical sucker evaginates and attaches to the wall of the small intestine. Development proceeds with the production of lateral suckers and the apical sucker gradually disintegrates giving the adolescent worm a pointed look which disappears in the adult.

There is only one genus in this family, but eight species have been described and all have been found in rodents. Little appears to be known of the pathogenicity of these worms, but in personal observations of laboratory infections between 1 and 25 worms are commonly found with no apparent ill effects. *C. pusilla* is not strictly host-specific and may spread to other rodents maintained in the same building. An apparently similar species has been isolated from colonies of rats and mice where they were maintained in contact with infected hamsters (personal observation). Another species found in the gerbil, *C. oranesis*, has been described by Joyeaux & Foley (1930) but has not been reported in laboratory mice or rats.

Family Anoplocephalidae

Another cyclophyllidean cestode which is only likely to be found in free-living monkeys and apes is *Bertiella studeri*. It is a very common parasite of a whole range of non-human primates (Cram 1928) (Fig. 3.31, page 85). The adult is a large worm situated in the small intestine and measures 10–30 cm in length × 1 cm wide. The segments are shallow, markedly wider than long. The scolex is sub-globose and unarmed. The eggs are thin-walled, have a crumpled appearance when viewed microscopically, and measure 38–45 μ (Tanaka *et al.* 1962). The onchosphere is surrounded by a pyriform inner shell with horn-like protrusions.

The eggs are passed out with the faeces and are ingested by free-living oribatid mites. The onchospheres migrate to the body cavity of the mite, where they develop into a typical cysticercoid larval stage, and the cycle is completed when the definitive host ingests the infected mite with its vegetable food. A related species, *Atriotaenia*, has been found in 23% of Columbian squirrel monkeys (Dunn 1968).

Order Pseudophyllidea

Members of the order Pseudophyllidea are parasites of reptiles and birds and carnivorous mammals, particularly those living close to fresh water lakes. The adult parasite therefore is not found in the laboratory under normal circumstances, although the larval stages are found in wild primates brought into captivity.

They are inhabitants of the small intestine of the definitive host, large in size with weak, slit-like bothria and no hooks on the scolex. The proglottids are wider than long and the eggs are shed from the genital pore. These are light brown in colour and thin-shelled with a small inconspicuous operculum. They measure 59–71 × 42–49 μ and when passed into the fresh water pond or lake release a ciliated coracidium which is next ingested by a copepod. The larva migrates to the haemocoele where it develops over a period of 14–18 days into a mature procercoid. The next stage is ingestion of the copepod by a fish where the procercoid migrates into the connective tissue and transforms to the infective stage, the pleurocercoid. The definitive host is infected by ingesting the raw intermediate host. Under ideal circumstances the prepatent period from egg to adult can take as little as 5–6 weeks.

The name *Sparganum* is given to the unidentified pleurocercoid larvae found in the connective tissues of a range of amphibia, birds and mammals including primates. They can be found in any region of the body and their migration may result in inflammation and tissue necrosis. They are typically white, ribbon-like organisms varying greatly in size and movement and usually the anterior region is inverted into the body. In old infections only degenerating and calcified portions of the worm may be found (Myers & Kuntz 1967; Kuntz *et al.* 1970). There have been a number of reports of spargenosis in non-human primates, including baboons (Kuntz & Myers 1967; Worms 1967) and African green monkeys (Moreton 1969).

CLASS TREMATODA (Flukes)

The trematodes or flukes are a completely parasitic group within the Phylum Platyhelminthes. Because of their need for one or more intermediate hosts, they are rarely found in animals bred in captivity. As with other complex parasites they can still be found in wild-caught species, particularly non-human primates.

Fasciola hepatica is a fluke commonly found in the bile ducts of sheep in moist areas where the intermediate host, a snail—usually *Lymnaea* species—is abundant. It has, however, been found in the liver of some rodents, such as the guinea pig, which have become infected when fed contaminated greenstuffs. Many species are vulnerable to infection, including carnivores, non-human primates and man, where the organisms have been known to do considerable damage during their migration.

The fully grown adult fluke is dorsoventrally flattened, leaf-like in shape and has a spiny surface. It measures 30 × 13 mm and has two well-developed suckers in the anterior region on the dorsal surface. The eggs are shed into the bile duct from a genital opening on the dorsal surface and are carried to the exterior in the host's faeces. The egg is large, measuring 130–150 × 63–90 μ, operculated and contains an undifferentiated mass of cells. The first larval stage is a very active, torpedo-shaped, ciliated larva known as miracidium. This requires moisture to hatch, following which it swims about seeking a suitable snail host. Its well-developed penetration mechanism allows it to enter the fleshy region of the foot, and once in the tissues it begins a series of multiplicative stages starting with what is known as a sporocyst. This is a sac-like body within which the next stage, the rediae, are developed. When these cigar-shaped bodies emerge they settle in another portion of the foot and each may produce either another redia generation or the final larval stage, the cercaria. This stage is very active, with a somewhat bullet-shaped anterior portion having two suckers and a slender tail. Under suitably moist conditions it migrates out of the snail and swims towards vegetation where it settles, rounds up, loses its tail and forms a resistant cyst around itself. This is the infective stage for the definitive host and is known as the metacercaria.

When ingested by the sheep during grazing, the cercaria emerges into the small intestine, migrates through the mucosal wall into the body cavity and finds its way to the liver. Once established in a bile duct it attaches to the wall and is fully mature at three months post-infection. As with the majority of flukes it is hermaphroditic, having extensively branched gonads, and is geared for massive egg production.

Many of the parasitic flukes of man have been reported in non-human primates. *Opisthorchis*, for example, is a small liver fluke of man which has been reported in cynomolgus monkeys (Habermann & Williams 1958) as well as in carnivores such as the cat and dog. This parasite requires a second intermediate host, a fish, where the metacercariae are to be found in the muscle and are ingested with the raw meat. The adults are small, measuring 7–18 mm long and 1.5–3.0 mm wide, as are the eggs (26–32 × 11–15 μ). The shell is very thick and dark in colour, but most distinctive is the striking convex operculum and the egg is laid with a fully developed miracidium within.

Another occasional inhabitant of primates is *Paragonimus westermanii*, a troglotrematid, one of the lung flukes. This parasite also requires a second intermediate host, the metacercariae in this case being found in a crustacean. The adult worms are bean-shaped, dark in colour and are usually found in pairs in a nodule in the lungs. They measure 7.5–12.0 mm in length, 4–6 mm wide and 3.5–5.0 mm thick. The eggs, which are pale in colour with a flattened operculum at one end, measure $80–118 \times 48–60\,\mu$. They are released into the air passages, are coughed up and swallowed and passed in the faeces. Development progresses through the snail intermediate host to a crustacean, and the definitive host is infected when the crab or crayfish containing the metacercariae is eaten raw or, in the case of humans, partly cooked.

There are many species of fluke which are occasional parasites of carnivores or non-human primates, but the most widespread are members of the Schistosomatidae. These are known as the blood flukes and in this group the sexes are separate. *Schistosoma japonicum* is found in the portal veins of man and domestic animals in Eastern Asia; it is truly zoonotic and is a naturally occurring infection of rats and other species. *S. mansoni* is found in the mesenteric veins of man and non-human primates in Africa, South America and the West Indies; and *S. haematobium* in the pelvic and mesenteric veins of non-human primates and man in Africa, Western Asia and Southern Europe.

None of these is of any pathological importance in non-human primates and they are usually only found at post-mortem examination (Kuntz & Myers 1967). The adults are found paired in the blood vessel, the cylindrical female enveloped by the lateral edges of the leaf-like male. The males measure 12–20 mm (*S. japonicum*) and 6.4–9.9 mm (*S. mansoni*); and the females 15–30 mm (*S. japonicum*) and 7.2–14.0 mm (*S. mansoni*). The eggs are not operculated and are distinguished further from other trematode eggs by the presence of spines which are diagnostic for the three species. It is these spines which allow the egg to work its way through the bladder or rectal wall. The inflammation which inevitably accompanies the passage of the egg, and which is caused by a toxic secretion of the egg, has been found to be a major contributory factor in ensuring that the egg is expelled (Doenoff *et al.* 1985; Vierra & Moraes-Santos 1987). At post-mortem examination eggs and adults of schistosomes have been found in squirrel monkeys and baboons (Fig. 3.32, page 86) recognizable by their distinctive shape and the terminal spine of the egg.

Other trematode species have been observed by the author in the small intestine (Fig. 3.33, page 87) of cynomolgus monkeys and may well be a species of *Dicrocoelium* which is common in non-human primates but usually associated with the gall bladder and bile ducts. What might well be a species of the oriental lung fluke *Paragonimus* has also appeared in sections, and occasionally *Paramphistomium* (Fig. 3.34, page 88) as well as *Gastrodiscus* (Fig. 3.35, page 88) have been recorded. A more complete list of the trematodes of non-human primates is given by Kuntz (1972).

PHYLUM ACANTHOCEPHALA
(Thorny-headed worms)

These are known as the thorny-headed worms because of their very distinctive, thorny proboscis. This is a large extrusable finger-like structure in the anterior region of the body used to anchor the worm firmly in the intestinal mucosa. They are sexually dimorphic and in order to complete their life cycle they need an arthropod intermediate host and sometimes a further one, to act as a transport host.

Order Oligocanthorynchiida

Moniliformis moniliformis is a parasite of rats, cosmopolitan in its distribution but found in larger numbers in the tropics or sub-tropics. It has been reported in other rodents but only rarely has it been found in man (Witenberg 1964), and once in the chimpanzee (Ruch 1959). They are large worms, the female measuring 10–32 cm and the male 6–8 cm. The proboscis is club-shaped, having 12–14 irregular rows of crescent-shaped hooks with which the worm can become firmly embedded in the wall of the mucosa, and the trunk has conspicuous pseudosegmentation. The eggs are large and oval in shape, measuring 90–125 × 50–62 μ. They have a lightly sculptured outer wall through which the developed embryo can be seen with its distinctive hooks at the anterior pole. The eggs are passed out with the faeces and then require to be eaten by a suitable intermediate host, usually a cockroach.

Once the acanthor larva emerges in the stomach of the insect, it burrows through the stomach wall using the embryonic hooks and into the haemocoele where it develops into an infected acanthella or cysticanth over a period of seven weeks. The definitive host becomes infected when it ingests a cockroach containing a mature acanthella. In the small intestine of the host, the cysticanth emerges from the cyst and attaches to the gut wall. The prepatent period for *Moniliformis* is 5–6 weeks. Light infections may cause very little clinical change but very heavy infections can result in ulceration of the mucosa.

The most pathogenic members of the Phylum Acanthocephala affecting laboratory primates, particularly New World monkeys, are members of the genus *Prosthenorchis—P. elegans* and *P. spirula* (Stunkard 1965). These parasites were an important cause of death when South American monkeys were transported in large numbers for research purposes. More rigorous controls on export imposed by many South American countries have resulted in species such as the marmoset being supplied almost entirely from laboratory-bred colonies.

Prosthenorchis is a cylindrical worm slightly curved in shape and with a globular proboscis armed with six spiral rows of 5–7 hooks (Fig. 3.36, page 88). The female measures 30–50 mm long and the male 20–33 mm (Machado-Filho 1950). The eggs measure 42–53 × 65–81 μ and contain an armed embryo as described above. The two species are very similar in all aspects except for the surface of their eggs, those of *P. elegans* appearing stippled and those of *P. spirula* being smooth. The trunk of *P. elegans* is slightly wrinkled and at the anterior junction of the proboscis and the trunk there is a distinctive collar of festoons not seen in any other species.

Prosthenorchis sp. can attach to any region of the intestine but are usually found in the terminal portion of the intestine from the ileum to the rectum. The proboscis is usually deeply embedded in the mucosa and causes extensive damage, often penetrating through the wall and causing peritonitis. At the point of attachment inflammation and ulceration occur and sometimes perforation of the intestine. The parasite has been known to enter the body cavity (Nelson *et al.* 1966).

It takes only a few of these to cause severe disease problems and in the marmoset, in particular, infection often results in death. Heavy infection is related to the level of cockroach infestation in the holding area. Heavy mortality due to this parasite is usually the result of the insanitary conditions under which the monkeys were maintained since capture and during transportation. *P. spirula* is less common than *P. elegans* but has been implicated in epizootics within groups of captive lemurs (Brumpt & Urbain 1938; Brumpt *et al.* 1939). The life cycle is identical to that of *P. elegans*.

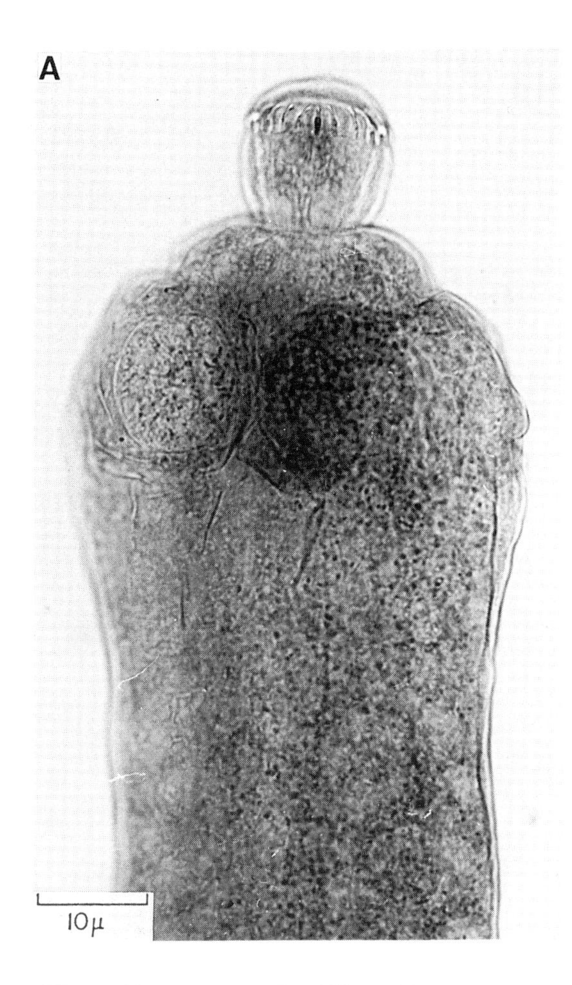

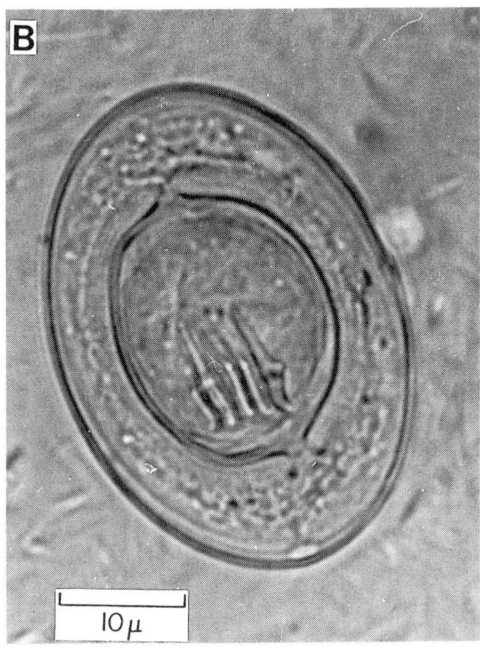

Fig. 3.23 *Hymenolepis nana*. **A**: Scolex of the adult worm with rostellum extended. **B**: Distinctive egg containing onchosphere. **C**: Cysticercoid within the villus of a mouse.

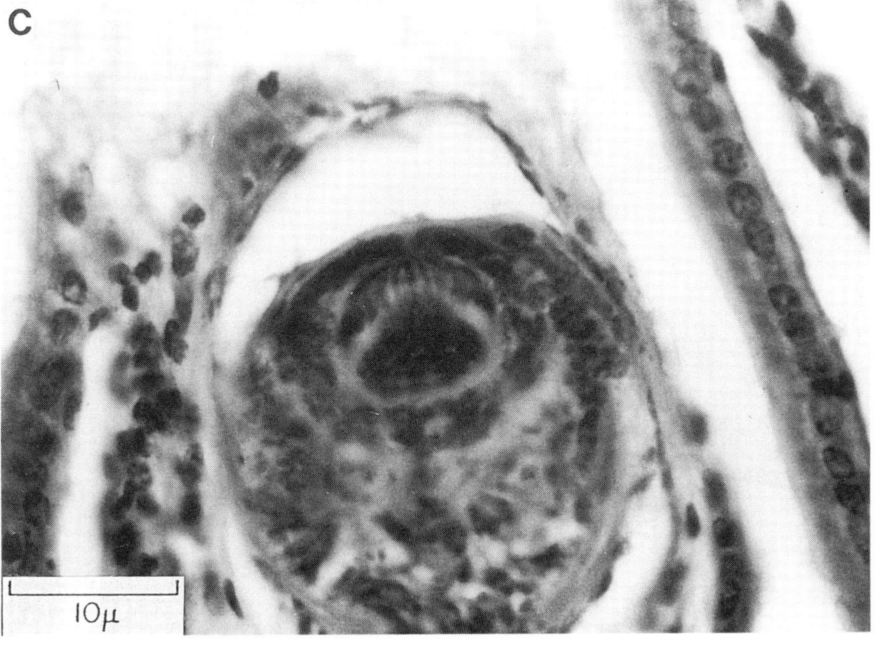

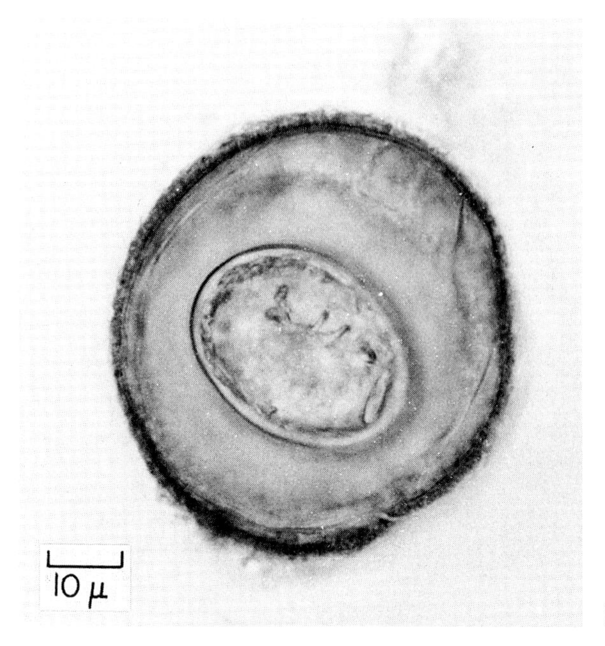

Fig. 3.24 Egg of *Hymenolepis diminuta.*

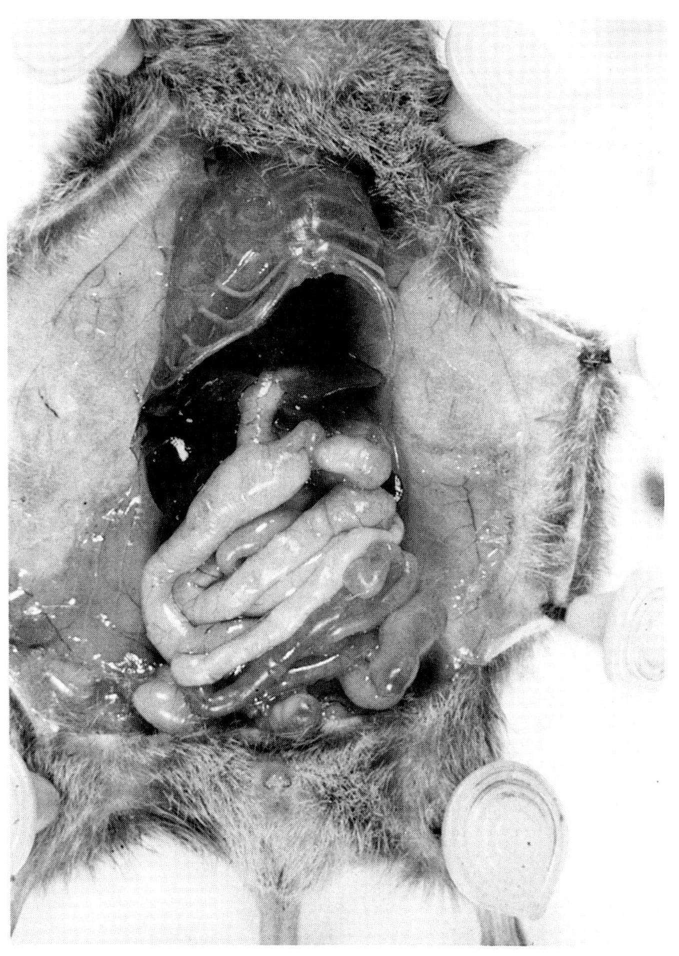

Fig. 3.25 Distended duodenum and bile ducts of a mouse infected with *Hymenolepis microstoma.*

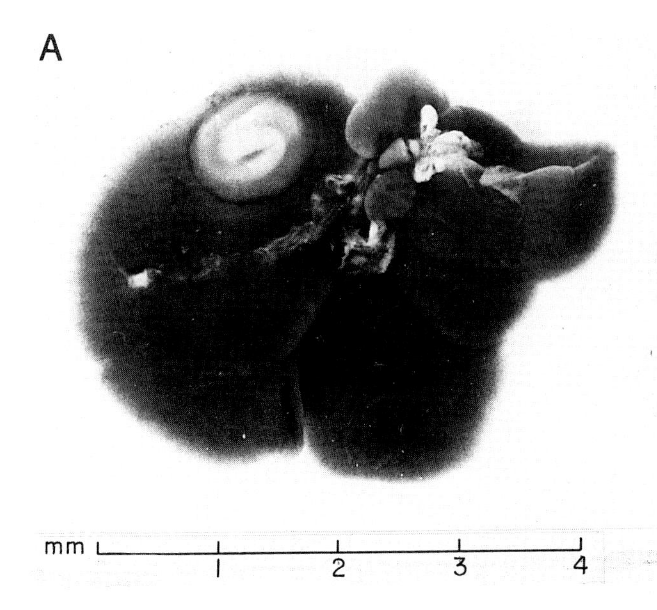

Fig. 3.26 *Cysticercus fasciolaris* larva from the liver of a mouse. **A**: Typical mature cyst just below the liver capsule. **B**: Section through cyst showing pseudo-segments and small bladder. **C**: Larva dissected from cyst.

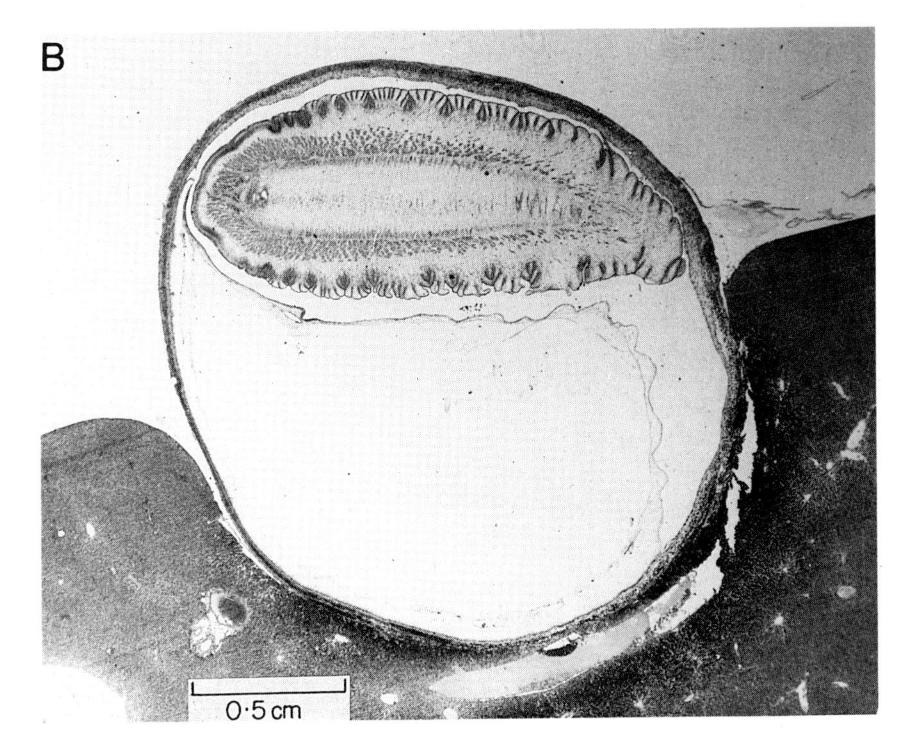

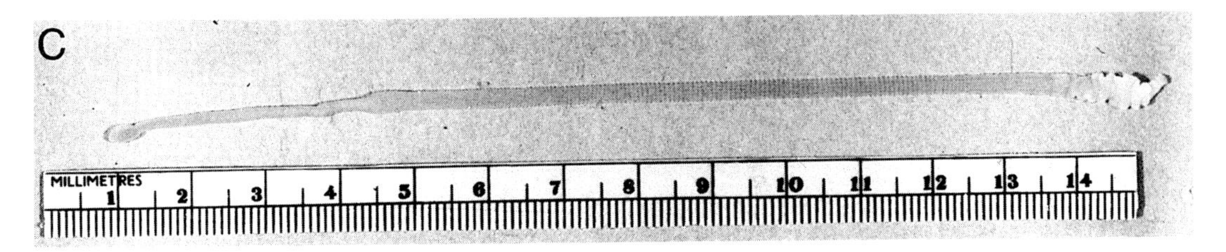

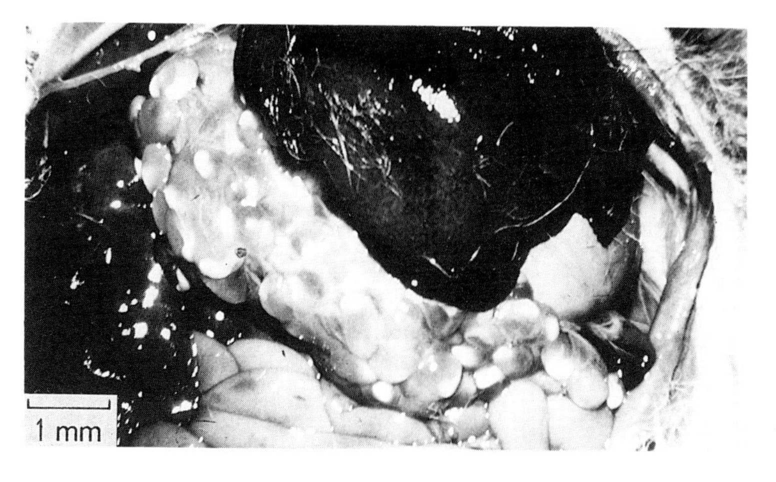

Fig. 3.27 Mesentery of a rabbit infected with *Cysticercus pisiformis*, the larval stage of *Taenia pisiformis*.

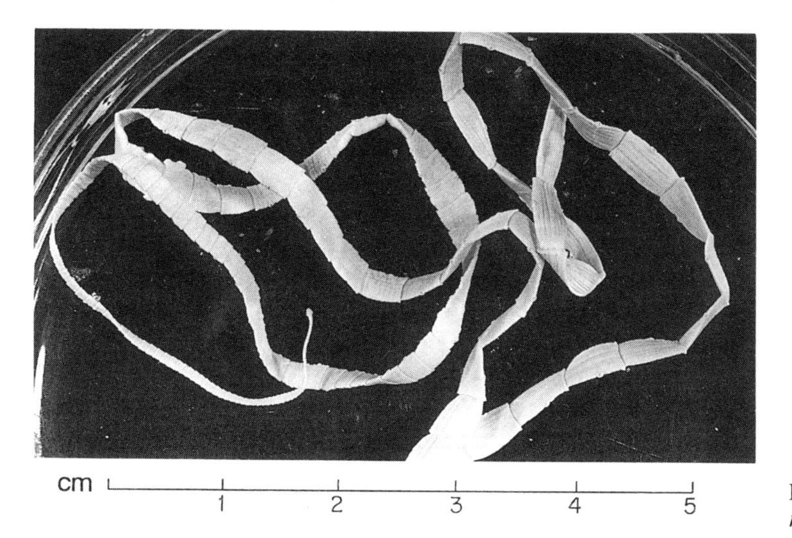

Fig. 3.28 Adult *Taenia multiceps*.

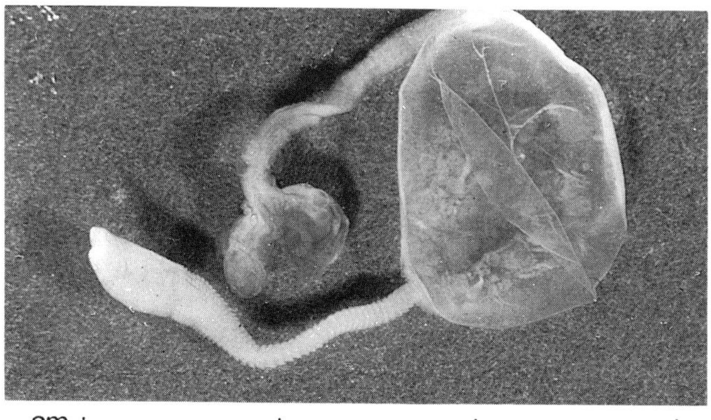

Fig. 3.29 Unidentified tetraphyllidean cestode from beneath the skin of a *Macaca mulatta* (Specimen originating from Dr F Taffs, NIBSC, Potters Bar).

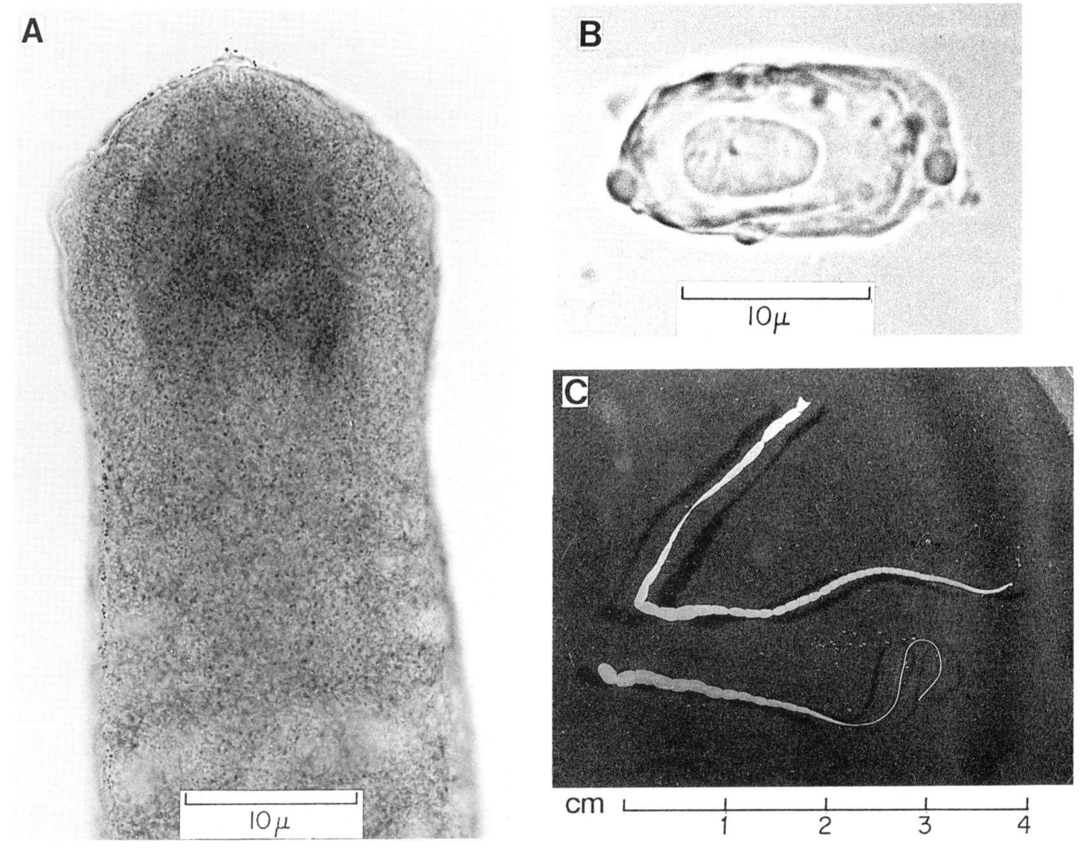

Fig. 3.30 *Cataenotaenia pusilla* from a golden hamster. **A**: Scolex of adult worm. **B**: Egg with six-hooked onchosphere. **C**: Adult worms showing proglottids longer than wide.

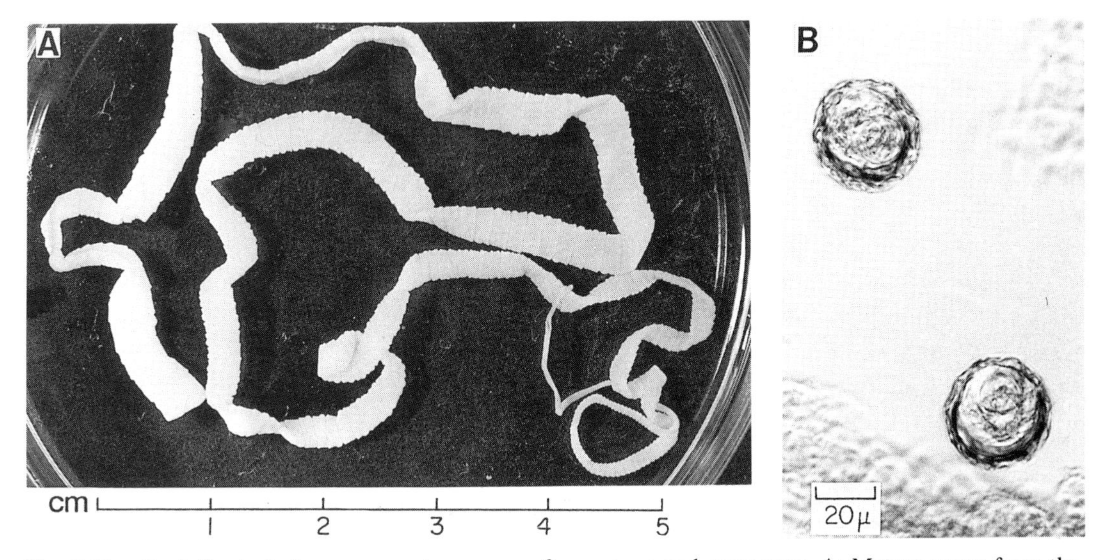

Fig. 3.31 *Bertiella studeri*, a common tapeworm of macaques and great apes. **A**: Mature worm from the small intestine showing segments broader than long. **B**: Egg containing six-hooked onchospheres.

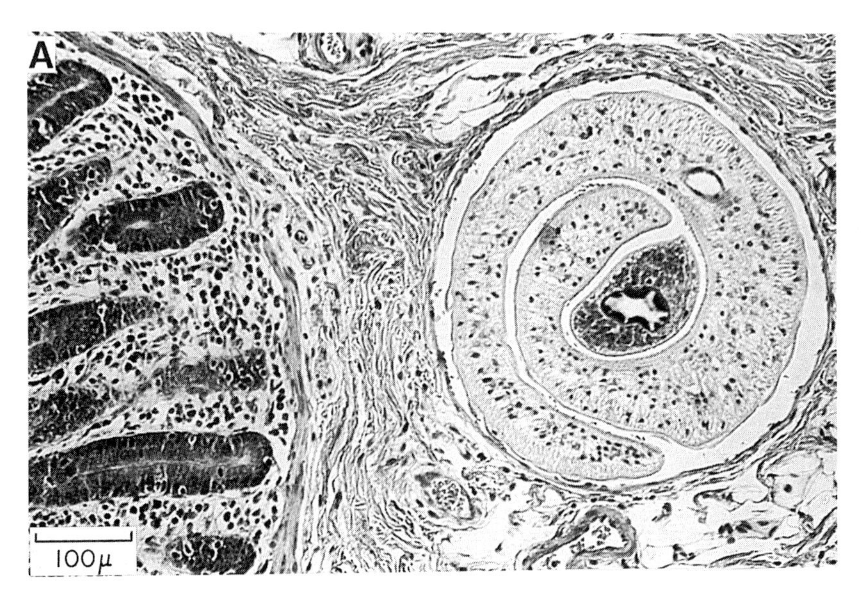

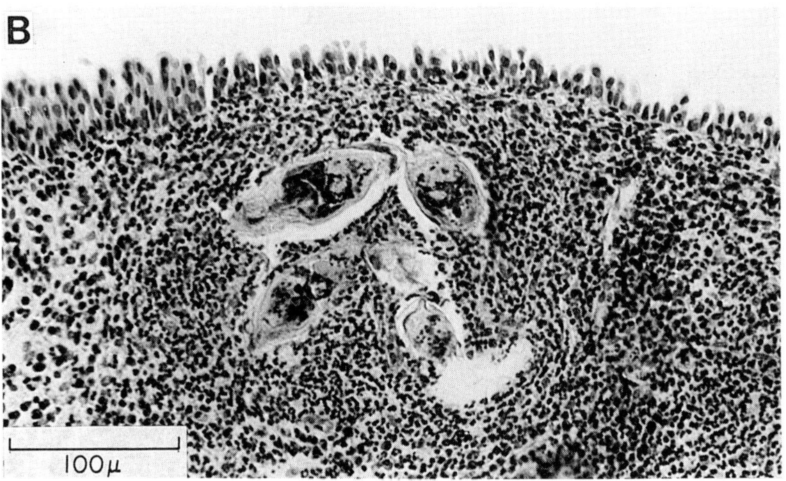

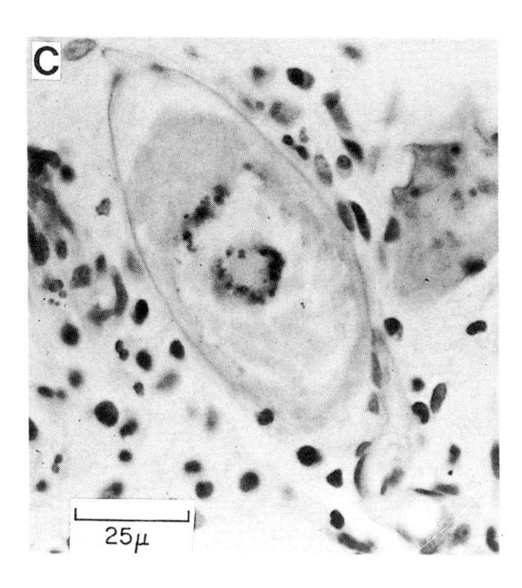

Fig. 3.32 *Schistosoma* sp. **A**: Pair of schistosomes in blood vessels surrounding the large intestine of a squirrel monkey. **B**: *Schistosoma* eggs surrounded by inflammatory reaction in the bladder of a baboon. **C**: Higher magnification of the egg showing a distinctive terminal spine.

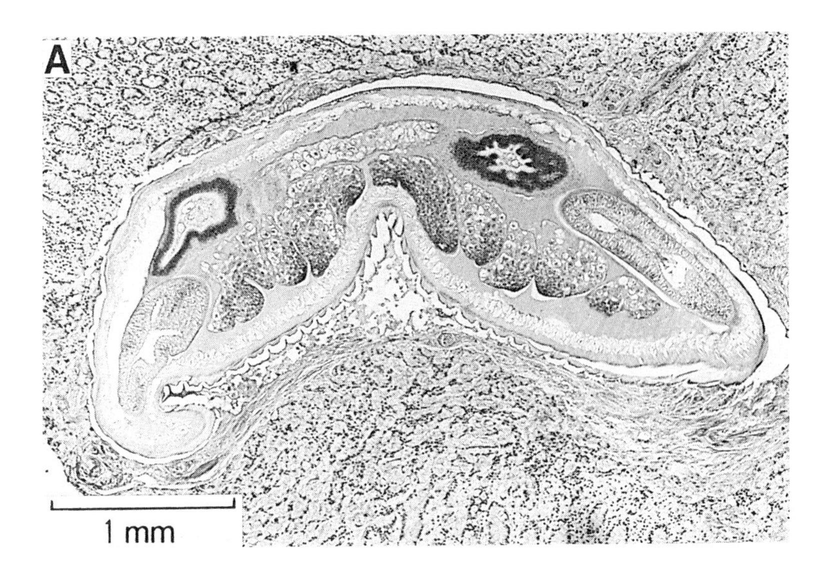

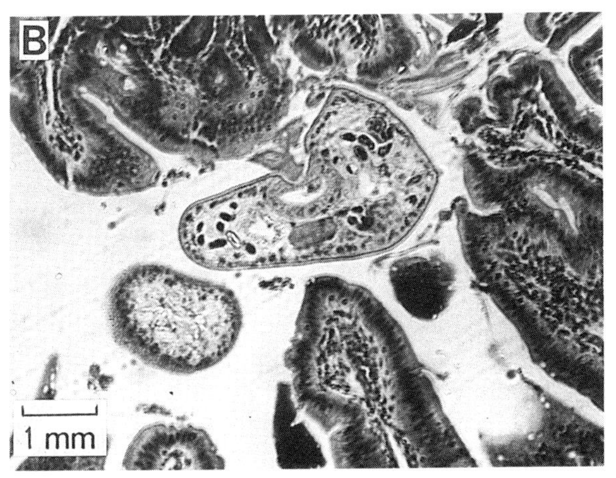

Fig. 3.33 A & B: Some sections through flukes in the small intestine of a cynomolgus monkey and thought to be *Dicrocoelium* sp. **C**: Section through two flukes in the liver of a marmoset, also thought to be *Dicrocoelium* sp.

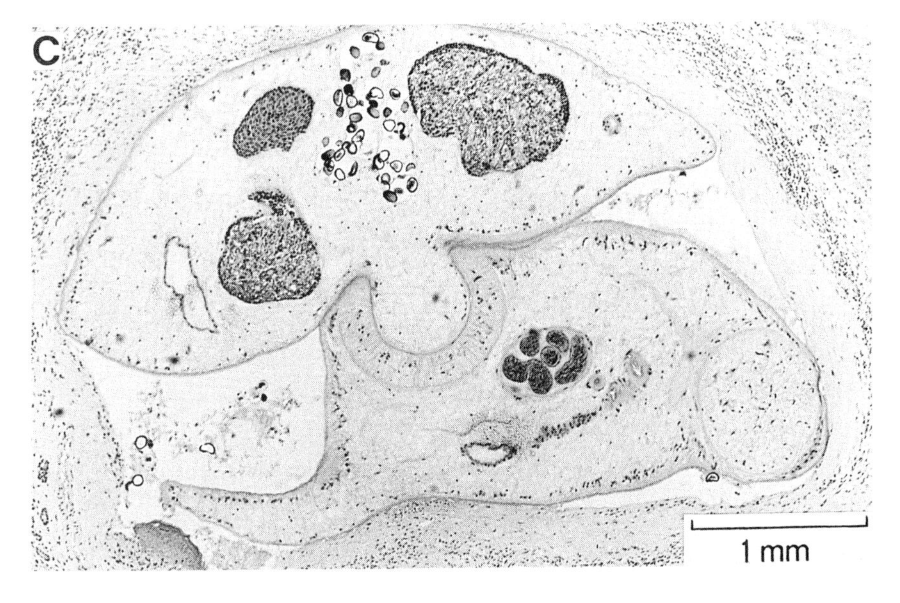

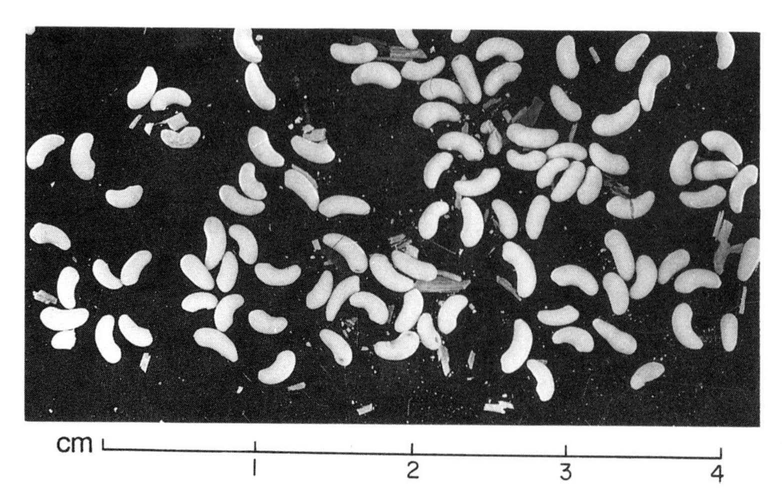

Fig. 3.34 *Paramphistomium microbothridium.*

Fig. 3.35 *Gastrodiscus* from the caecum of a macaque.

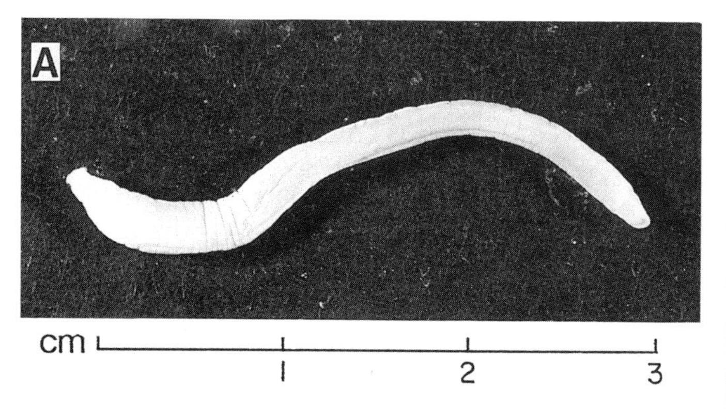

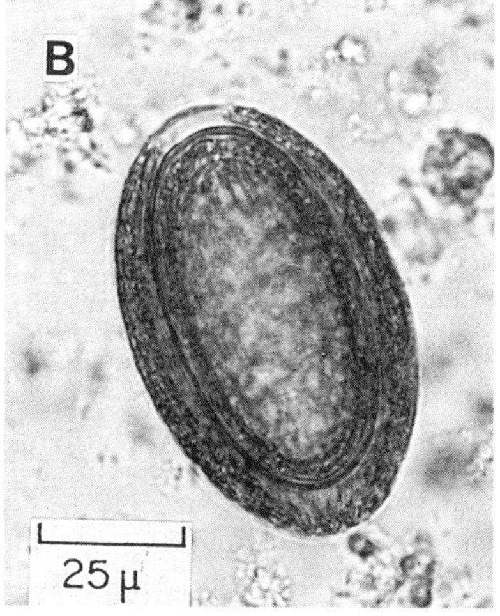

Fig. 3.36 *Prosthenorchis elegans*, from a squirrel monkey. **A**: Adult worm. **B**: Thick-walled egg with distinctive hooked acanthor.

PHYLUM NEMATODA
(Roundworms)

The nematodes, or roundworms, are among the most numerous endoparasites of laboratory animals and for this reason they will be considered here under their respective orders. The classification used is that drawn up by Levine (1968). There are in fact two classes and eleven orders; of these, five orders are considered here as being of importance.

All nematodes have a similar basic structure. They are cylindrical in shape with a simple digestive tract, firm cuticle and the sexes are usually separate. Variation appears in size differences and the presence or absence of appendages round the oral or anal apertures. Parasitic nematodes can be isolated from a wide variety of sites and the life cycle ranges between the very simple direct host-to-host of the pinworm, to the highly complex cycles found in the Spirurorida in which one or more intermediate hosts is essential for survival.

ORDER ASCARIDORIDA

Family Oxyuridae (Pinworms)

The family Oxyuridae comprises the pinworms, one of the most common parasites in laboratory rodents.

Syphacia obvelata (Fig. 3.37, page 96) is perhaps the most well known parasite of mice as it is the most difficult to eradicate from a conventional animal house. All stages of the worm are to be found in the caecum. The female measures 4–6 × 0.13 mm, and the smaller male 1.3–1.6 × 0.11–0.22 mm. The mouth is typically surrounded by three cushion-like simple lips and there is a club-shaped oesophagus and a very prominent oesophageal bulb. One of the common features of the pinworm is the presence of distinctive lateral extensions of the cuticle around the anterior region. In *Syphacia* these are simple flaps which taper to the side of the body over the bulb region. In this species the body of the female tapers to a long, slender point behind the anal opening. The posterior extremity of the male is bent slightly ventrally and a single prominent spicule is present. After fertilization the male dies and is passed out with the faeces. The gravid female, when ready, migrates to the anus approximately 12 days after fertilization and proceeds to lay eggs by protruding the anterior region through the host's anus and depositing the ova. The prepatent period is 15–16 days and after depositing the eggs, usually around 350 per worm, the female dries up and dies.

The eggs are thin-walled and flattened along one side and measure 100–142 × 30–40 μ. Embryonation is rapid, taking only 6–24 hours, after which the ova are infective to other mice which may ingest them. Once in the alimentary tract the eggs hatch and the larvae continue their way to the caecum where they moult through the fourth stage to sexually mature adults (Chan 1952; Oldham 1967; Flynn 1973).

Syphacia muris is an almost identical species found in rats but is apparently unable to infect mice except under circumstances of high challenge (Hussey 1957). In the author's experience, *S. obvelata* failed to infect two rooms of rats in an SPF building afflicted with an outbreak in mice. Rapid measures were taken to prevent spread, but it is interesting that although eight mouse rooms were affected, the two rat rooms were not (Owen & Turton 1979). Another species, *S. mesocriceti*, has been described in hamsters (Dick *et al.* 1973), again very similar in general shape and size to other members of the genus but having distinctly reduced lips around a triangular stoma. The species *S. obvelata* has been observed in the mongolian gerbil (*Meriones unguiculatus*) (Wightman *et al.* 1978*b*) and experiments showed that the parasite was easily transmitted between mice and gerbils.

Another naturally occurring parasite of the gerbil is the oxyurid *Dentostomella translucida*. This is a fairly large worm found in the small intestine, principally in the anterior third (Wightman *et al.* 1978*a*). Females measure 9.63–13.0 mm in length and males 6.14–13.14 mm, and the thick transparent cuticle has coarse striations. The anterior portion has small cervical inflations, lips are absent and the buccal cavity is armed with teeth (Schulz & Krepkorgorskaja 1932). One of the criteria for identification is the number of teeth in each sector; in the case of *D. translucida* there are five teeth. Details of the life cycle of the members of this genus are not known, though from the evidence collected by Wightman *et al.* (1978*a*) it is likely to be direct, and the prepatent period around 25–29 days. The eggs are found in the faeces and measure 115–140 × 50–62 μ.

The second most common pinworm found in the mouse is *Aspiculuris tetraptera* (Fig. 3.38, page 97) (Boecker 1953), a fairly small worm, the female measuring 2.6–4.7 × 0.19–0.25 mm and the male 2.26–3.2 × 0.15–0.17 mm. They are found in the lumen of the caecum and upper colon usually in a distinct tangle of mixed sexes. Typically there are three lips around the mouth, but anteriorly there are two distinctive flap-like cervical alae which end in the region of the posterior bulb and which give the head an arrow-like appearance. In both sexes the tail is blunt and conical in shape, and the male has no spicule. The eggs are voided in the faeces, usually in reponse to activity of the host (Phillipson 1974); they are oval in shape with a robust double wall and measure 90 × 40 μ. Infection is by ingestion of the ovum which is embryonated in 7 days and development proceeds as in other members of the family, the prepatent period taking 23 days.

All species of pinworm are regarded as having little deleterious effect on their hosts, producing no clinical signs even when present in large numbers. However, they have been thought to affect weight gain and growth rate and various disorders of the intestine have been attributed to *Syphacia* species (Flynn 1973). Although there has never been any conclusive proof that the pinworms by themselves elicit clinical signs, they are known to provoke a thymus-dependent immunological response in the host (Jacobson & Reed 1974). Groups of athymic (nude) mice and their normal litter-mates together with athymic mice implanted with thymus gland were all infected with *A. tetraptera* and *S. obvelata*. With increasing age, when it would be expected that the intact mouse would have a reduced burden, the athymic mouse burden continued to increase. The intensity of infection would therefore appear to depend on the type of host, as has already been postulated (Eaton 1972), and the innocuous reputation of the parasite would seem to be questionable, at least among the rodents.

Another apparently non-pathogenic oxyurid, *Passalurus ambiguus* (Fig. 3.39, page 98), is found in the lumen of the caecum of the rabbit and hare (Chandler 1926; Bull 1953). This is typical of its family, the female measuring 9–11 mm and the male 4–5 mm. Distinctive cervical alae produce a lobulated pattern at the side of the head and the mouth has three small lips. The oesophagus has a strongly marked oesophageal bulb with a prebulbar swelling. In the male the caudal alae are supported by small papillae and a single spicule is present. In both sexes the tail is long but in the female it becomes exceptionally long as it reaches maturation, when it can be 3.4–4.5 mm in length, and has strong cuticular striations along its length. As with other members of the family, the life cycle is direct. The egg, which is 95 × 103 × 43 μ in size and slightly flattened, is voided in the faeces and infection of the next host is by ingestion. It is a fairly common parasite of wild rabbits, less common in the laboratory and is not generally thought to be pathogenic. When present it is usually easily detected in young animals where it may be present in very large numbers. It is possible that these large numbers of worms could be responsible for gastric disturbances and there has been discussion as to their being responsible for mortality in young rabbits (Edgar 1933), though there have been no recent reports of actual instances.

The human pinworm, *Enterobius vermicularis*, has been reported in a variety of non-human primates (Ruch 1959). In some cases the parasite has produced what is thought to be a hypersensitivity reaction, as was seen in a chimpanzee where it was associated with severe polyposis (Toft *et al.* 1976). It is typical of the family, the female measuring 8.0–13.0 mm in length, the male 2.0–5.0 mm. The life cycle is direct, as in all others in the family, with the female laying the eggs directly onto the perianal skin of the host. The eggs measure 50–60 × 20–30 μ, are slightly flattened on one side, and take only six hours to mature. Transmission is by ingestion of the embryonated ova and the full cycle is complete in 15–28 days.

There are many different species of *Enterobius* to be found among primates and within their natural host they appear to remain relatively non-pathogenic. To a certain extent the irritation of the active females emerging to lay their eggs can result in restlessness, scratching and perianal pruritis, although *E. anthropopitheci* (Fig. 3.40, page 98), found in the chimpanzee, has been shown to produce some internal tissue damage in its natural host.

According to Inglis (1961), *Enterobius* and *Trypanoxyuris* are placed together in the subfamily Oxyurinae and are distinguishable by the marked enlargement of the male caudal alae and their elongated supporting papillae. This classification does not accord with that of Levine (1968) which is followed here, but seems to re-emphasize the similarity between these two species. *Trypanoxyuris* (Fig. 3.41, page 99) has been described in many species of tamarin and marmoset.

Families Heterakidae and Ascarididae

These are small or medium worms with the typical oesophagus ending in a muscular bulb. The mouth usually has three lips and lacks any sort of vestibule. The male always has a well-defined chitinous sucker in the perianal region with slender chitinous spicules.

Heterakis spumosa (Fig. 3.42, page 99) is a relatively common parasite of the wild rat but is rarely found in the laboratory. The adults are found in the caecum. The female is a robust worm measuring 7–13 mm in length and 740 μ in width, with strong lateral extensions of the cuticle in the cervical region. Below the chitinous sucker the male tail is provided with caudal alae supported by ten pairs of papillae and the two long spicules are subequal. The male is shorter and more slender than the female, measuring 6–10 mm long and 260 μ wide.

The eggs have a very thick mamillated shell and measure 55–60 × 40–55 μ. They are passed in the faeces and become infective within 14 days if the temperature and humidity are suitable. Following the host's ingestion of the embryonated eggs, the larvae hatch in the small intestine and make their way towards the caecum where they can be found 72 hours after infection. Development to the adult takes place in the caecum but the worms migrate to the upper colon at the time of maturity. Eggs first appear in the faeces 26–47 days later (Flynn 1973; Oldham 1967).

The only member of the family Heterakidae to be found in laboratory animals is the so-called 'pinworm' of the guinea pig, *Paraspidodera uncinata* (Fig. 3.43, page 100) (Porter & Otto 1934). Again this is common only in guinea pigs kept in outside runs and is not usually found in caged animals. It is a medium-sized robust worm, the female measuring

16 mm in length by 400 μ in diameter, and the male 11 mm × 300 μ. The male has no caudal alae but the chitinous ring is present and short, stout, subequal spicules. In both sexes the mouth is surrounded by three distinctive cushion-like lips. Cervical alae are absent but the oesophagus swells gradually to a muscular posterior bulb. The eggs are very similar in general appearance to those of *H. gallinarum*, being ellipsoidal in shape, thick-shelled and measuring 43 × 31 μ. Infection of the host is by ingestion of the embryonated egg and mature worms are present in the caecum 6–8 weeks after infection. In neither of these species has there ever been any sign of pathological effects on the host.

Members of the subfamily Subuluridae (Fig. 3.44, pages 100–101) are to be found in a variety of primates; for example *Subulura andersoni* in the squirrel monkey, *S. jacchi* in *Callithrix jacchus* the common marmoset, and *S. distans* in *Cercopithecus* sp. This species has also been reported as spreading to the mandrill (Desportes & Roth 1943) and African green monkey (Canavan 1929) in zoos. These are small worms, the female measuring approximately 12 mm in length and the male 7 mm. Lateral extensions of the cuticle are usually distinctive. A small vestibule is present with three strongly chitinized teeth with sharp points at the base. The perianal sucker of the male is fusiform and set in front of the cloaca; spicules and longitudinal rows of papillae typical of the group are present, together with long slender spicules varying in shape according to species.

Ascaris suum (Fig. 3.45, page 101), an ascarid indistinguishable from *Ascaris lumbricoides* of man, has been reported from a wide variety of monkeys and apes. Dunn & Greer (1962), Baylis & Daubney (1922) and Orihel (1968) succeeded in experimentally infecting chimpanzees with human *A. lumbricoides* and found that the life cycle is similar to that in man, but that on the whole the animals were not uniformly susceptible to infection. Ascarids are large worms, the female measuring 20–40 cm, the males 15–30 cm. The relatively undistinguished cylindrical body has at the anterior end three lips around the mouth. The oval eggs are thick-walled and remarkably resistant. They measure 50–75 × 40–50 μ and have been known to survive for many years in suitable conditions. The embryonated egg ingested by the host hatches and the larva proceeds to migrate through the body wall to the liver, thence to the lungs where the fourth-stage larva breaks through into the lungs and travels up the air passages to the pharynx from whence it is swallowed back into the alimentary tract. Although fairly rare and innocuous in most apes, deaths in chimpanzees have been attributed to heavy infections of these worms and to their migration into other organs.

ORDER SPIRURORIDA

These worms are not found in laboratory-bred animals, but they are common parasites of the stomach and small intestine of wild-caught Old and New World monkeys. They are medium-sized worms and vary morphologically within the group, often with marked structural and size differences between the sexes. The male is smaller than the female, has no bursa but may have caudal alae. The tail is coiled and the spicules are of unequal size and shape. The eggs have a very characteristic thick-walled transparent shell and are embryonated when laid. The intermediate host is usually a scavenging arthropod such as a cockroach, and infection of the definitive host is by ingestion of the secondary host. Moderate infections seem to be tolerated with little or no clinical signs, but heavy infestations result in extensive gastric hyperplasia and can cause death if the stomach wall becomes perforated.

A rather large *Protospirura* (Fig. 3.46, page 102), *P. muricola*, was recorded by Foster & Johnson (1939) in a colony of *Cebus capucinus*. From the necropsy of the first two monkeys investigated, they collected 786 and 502 worms respectively. Spread was rapid in the colony and the worms were so numerous that when fixed they formed a cast of the organs. In the latter case little damage had been done to the stomach tissues, but sheer numbers of worms had caused mechanical blockage.

Physaloptera sp. are also known to cause mortality due to gastric hyperplasia, gastritis and colitis. They are found embedded in the stomach lining in a range of Old and New World monkeys. *P. tumefaciens* (Fig. 3.47, page 102) is a common species found in Asia where it has been found in 27.4% of cynomolgus monkeys (Hashimoto & Honjo 1966); *P. dilatata* has been found in capuchins, woolly monkeys and marmosets in South America. Only when present in large numbers do they cause gastric disturbance.

The eggs are oval measuring 50–70 × 25–35 μ, typically thick-walled and embryonated when passed and are easily found in the faeces. The adult worms are robust and similar in shape to ascarids (Levine 1968). They attach firmly to the mucosa. They have two simple triangular lips each armed with teeth and the cuticle is reflected over the lips to form a cephalic colerette. Their most characteristic feature is several rows of oval cuticular bosses at the anterior extremity. In the male the spicules are unequal and the alae asymmetrical, meeting in front of the anus (York & Maplestone 1933).

Over a dozen species of *Physaloptera* have been described from Old and New World monkeys. When present in large numbers they can cause gastric hyperplasia. Heavy infections have been described in the Javanese macaque (Finkeldy 1931) and fatal infections due to colitis and gastritis (Fox 1927, 1930).

Although *Gongylonema* sp. are generally considered to be parasites of birds, many species have been described in both Old and New World monkeys (Fig. 3.48, page 102). They are parasites of the mucosal layer of the oesophagus, but have been recovered from the bronchi and stomach and the buccal cavity (Gebauer 1933; Orihel & Seibold 1972). The worms are filariform in shape, the female measuring 5–25 × 15–55 mm, and migrate through the epithelium forming tunnels. The eggs, measuring 50–70 × 25–35 μ, find their way via the tunnels into the lumen and are voided with the faeces. They mature to infective larvae after a couple of days and the larva hatches when consumed by an arthropod host where it develops in the tissues to the infective stage. Infection of the definitive host is by ingestion of the arthropod containing the infective larva.

Trichospirura leptostoma is a thelazid spiruroid, a tiny worm found in the pancreatic ducts of the squirrel monkey (Smith & Chitwood 1967; Smith & Levy 1969); when present in very heavy numbers, it causes chronic pancreatitis. These are long, slender worms, the female measuring 20 mm and the male 15–20 mm. Infection is usually limited to 10–12 in number, but as many as 31 specimens have been found in the ducts. The eggs measure 50 × 25 μ, are typically thick-shelled, contain a larva and are found in the faeces.

Other thelazid nematodes common in the macaque are *Streptopharagus armatus* and *S. pigmentatus* (Fig. 3.49, page 103). They are found in the stomach wall, the female measuring 47–48 mm and the male 30–32 mm. The hexagonal mouth has small trilobed lateral lips and an asymmetrically inflated cervical cuticle. A number of small teeth project into the buccal cavity which is continuous with the very distinctive vestibule. Half way along its length the vestibule is twisted into a half spiral. The tail of the male is spiral, has broad, strong caudal alae and the spicules are unequal and dissimilar. The eggs are asymmetrical, measuring 28–38 × 17–22 μ, and typically are thick-shelled and embryonated when passed in the faeces.

Superfamily Filariicae

These are long, slender nematodes varying in length from a few centimetres to 30 cm and more; there is a marked difference in size between the sexes, the females always being much longer than the males. They are found in body cavities, subcutaneously in connective tissue or in lymph ducts or organs.

There are no representatives in laboratory-bred rodents, although there are plenty in the wild. *Litomosoides carinii* of the cotton rat was for many years used as a model for human filariasis, and many rodents such as the gerbil (*Meriones unguiculatus*) and the multimammate rat (*Mastomys natalensis*) are susceptible to infection with human species of filarial worms such as *Brugia malayi* and the laboratory models of the cat species *B. patei* and *B. pahangi* (Laing *et al.* 1960).

Filarial worms are most likely to be found in imported monkeys, usually New World species. Of the few found in Old World species, the most common are the heartworms, *Dirofilaria* sp. (Fig. 3.50, page 104), found in both African and Asian monkeys (Webber 1955*a*, *b*; Jackson 1969). Some, like *D. corynodes*, inhabit subcutaneous tissues and the slender first larval stage, the microfilariae, have no sheath and can be found in the peripheral blood at night.

Transmission is by means of a biting insect which acts as the intermediate host. Larvae in the ingested blood migrate through the gut wall of the insect and into the muscles of the thorax where they develop and moult through second to third generation over a period of 7–10 days. When mature, these third-stage larvae migrate to the head and into the mouthparts, and are released into a new host at the time of the next feed. They enter the host tissues and migrate to the favoured site, which could be in the skin or the lymphatics.

The prepatent period, during which there is a fourth moult to the adult stage, is around 30–60 days for most species, but patency in human *Onchocerca* infection has been known to take as long as 11–12 months following exposure. The longevity of the circulating microfilariae is calculated to be at least 100 days but is likely to be more in many host species.

Animals imported from South America are most likely to harbour a range of filarial worms. Two types of microfilariae are likely to be found in the blood (Eberhard 1978): *Dipetalonema* sp. and *Tetrapetalonema* sp. (Fig. 3.51, page 104) (Esslinger 1962, 1979). They may be present in very large numbers but appear to have little detrimental effect on the monkey. *D. gracile* and *D. caudispina*, however, are found in the body cavities and when present in large numbers can give rise to inflammation. The author has observed marmosets kept under laboratory conditions for many years which at necropsy were found to be harbouring large populations of *Dipetalonema* which had caused extensive adhesions in the abdominal cavity. The worms situated subcutaneously produced no apparent pathogenic effect. There are many species in South American monkeys in which the life cycle is very poorly understood and, judging from the range of microfilariae found in the blood, it is usual for at least two species to be present.

Other filarial worms associated with the great apes are zoonotic or closely related to the species found in man; for example, *Loa loa* in the gorilla, drill and baboon. The adults of these live in the subcutaneous tissues but severe pathology has not been caused by these worms. Some effect of the presence of the worm can be seen in the drill where the spleen has been described as 'nodulant' due to the death of microfilariae within the organ (Duke 1960).

The human species of *Brugia malayi* has been isolated in macaques. The adults of this species live in the lymphatic or perilymphatic tissues of their hosts and the sheathed microfilariae are found in the peripheral blood. Another species, *B. pahangi*, has been reported in humans but it is usually found exclusively in animals and has been found in a range of Asian monkeys. Although they are to be found in the same location as in man, no clinical signs of infection have been reported from either *B. pahangi* or *B. malayi* infection in primates.

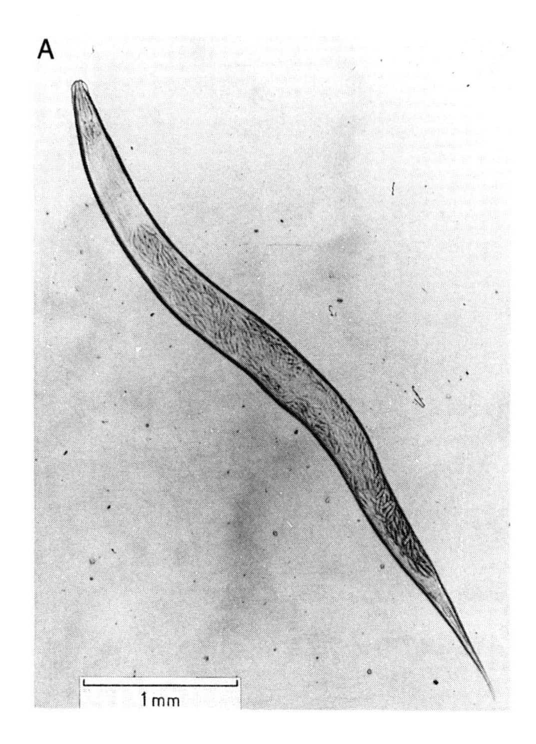

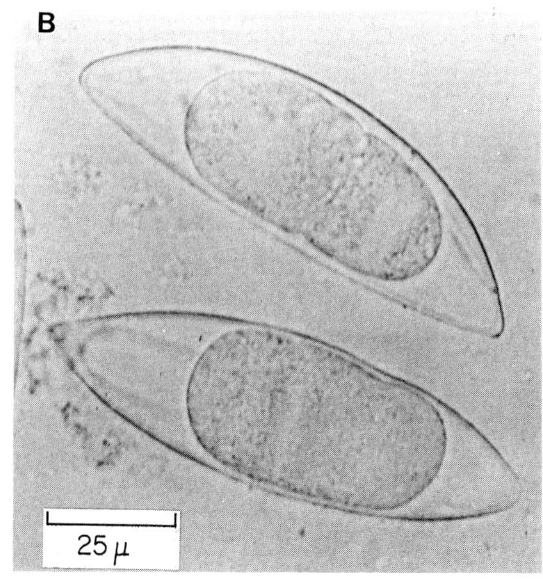

Fig. 3.37 *Syphacia obvelata*, 'pinworm' from a mouse. **A**: Adult female packed with eggs. **B**: Eggs. **C**: Head showing cushion-like lips. **D**: Slender pointed tail of adult female.

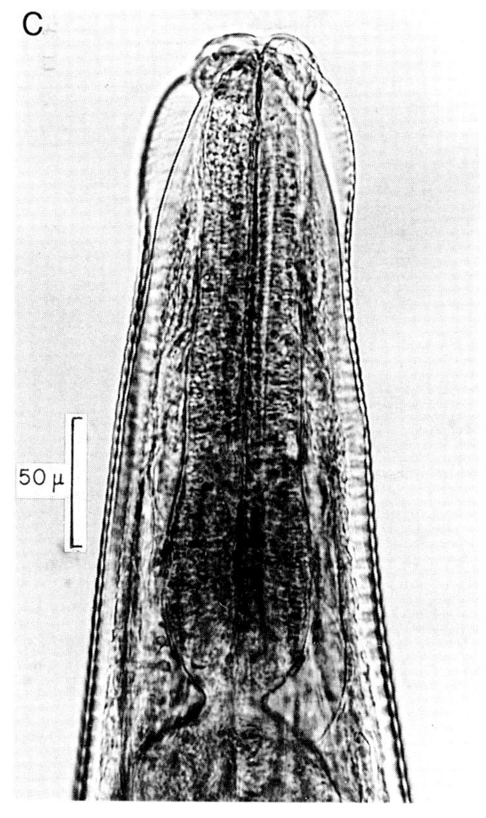

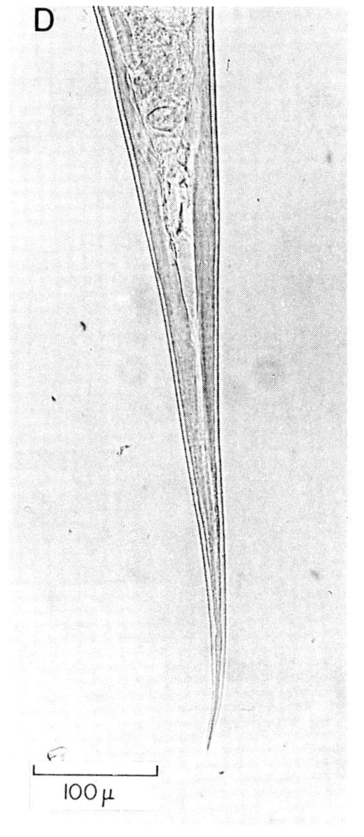

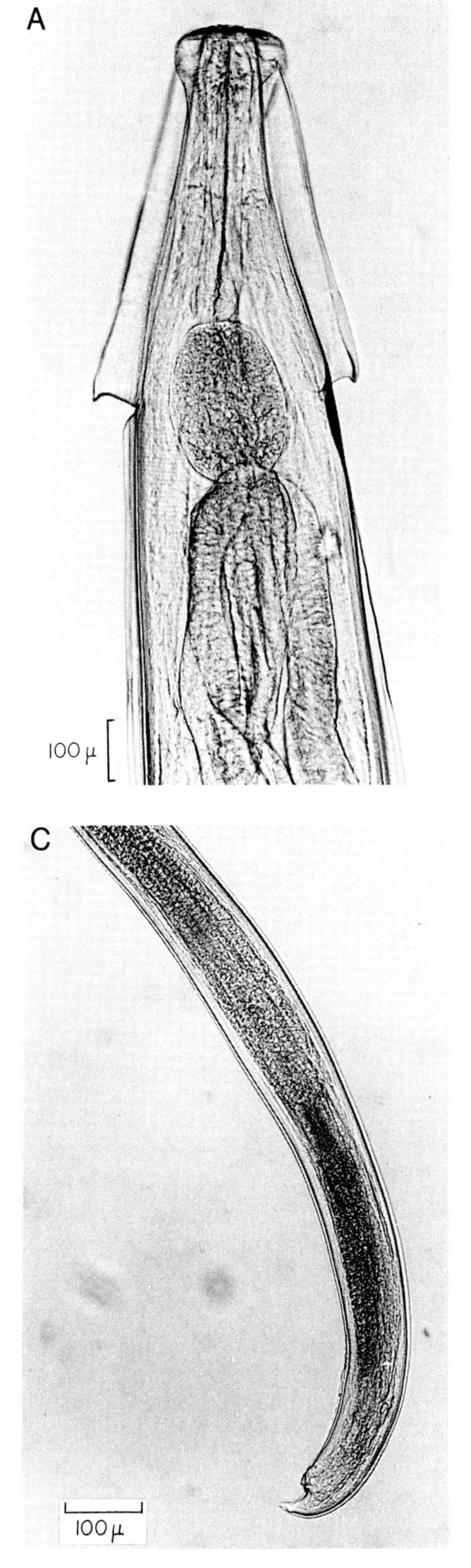

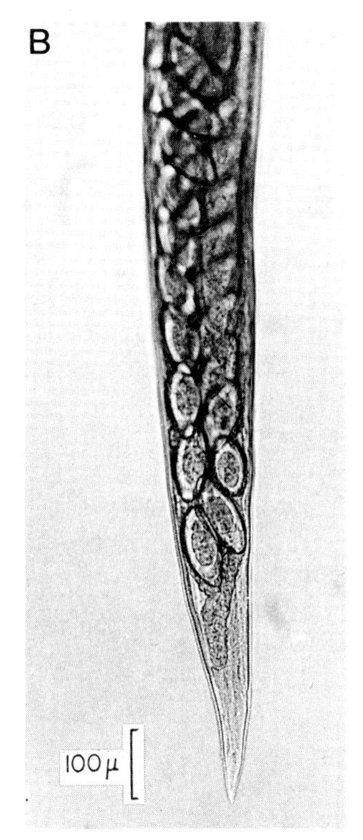

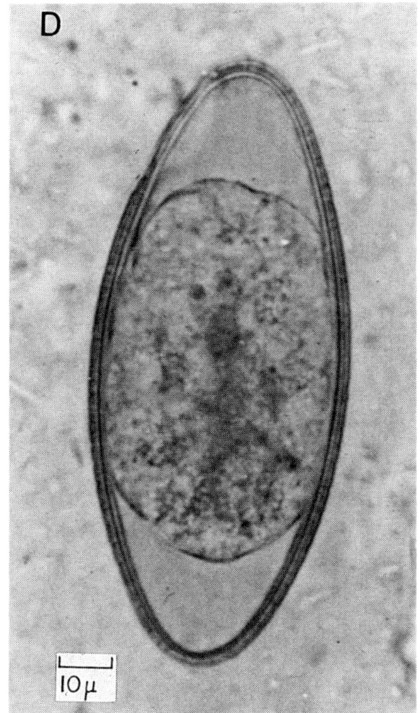

Fig. 3.38 *Aspiculuris tetraptera*, 'pinworm' from a mouse. **A**: Head showing expanded cervical alae giving an arrow-shaped appearance. **B**: Blunt-ended tail of female. **C**: Tail of male dorsoventrally curved with a short spicule. **D**: Egg.

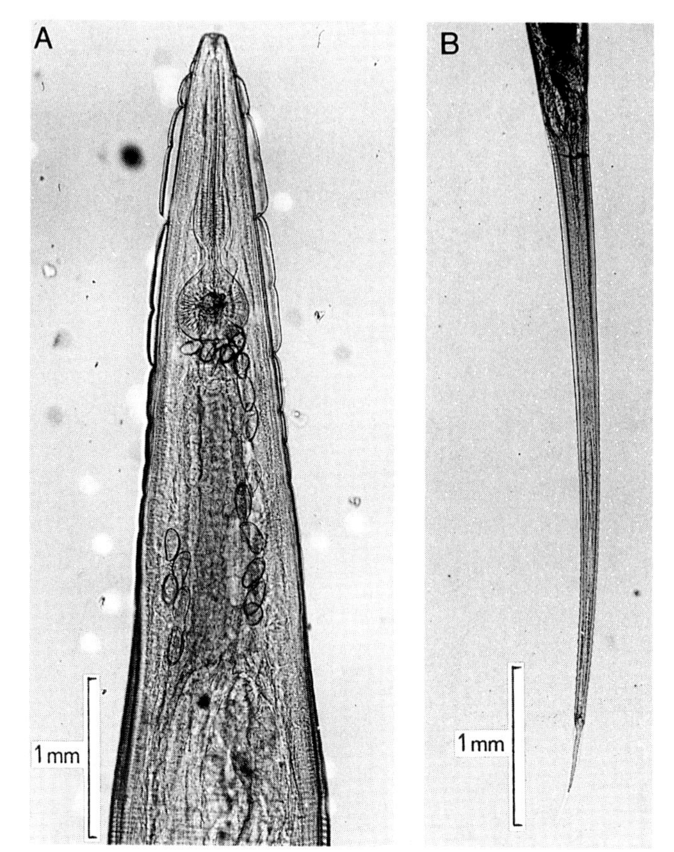

Fig. 3.39 *Passalurus ambiguus*, 'pinworm' from a rabbit. **A**: Head of female showing narrow alae. **B**: Long slender tail of female with fine annulations.

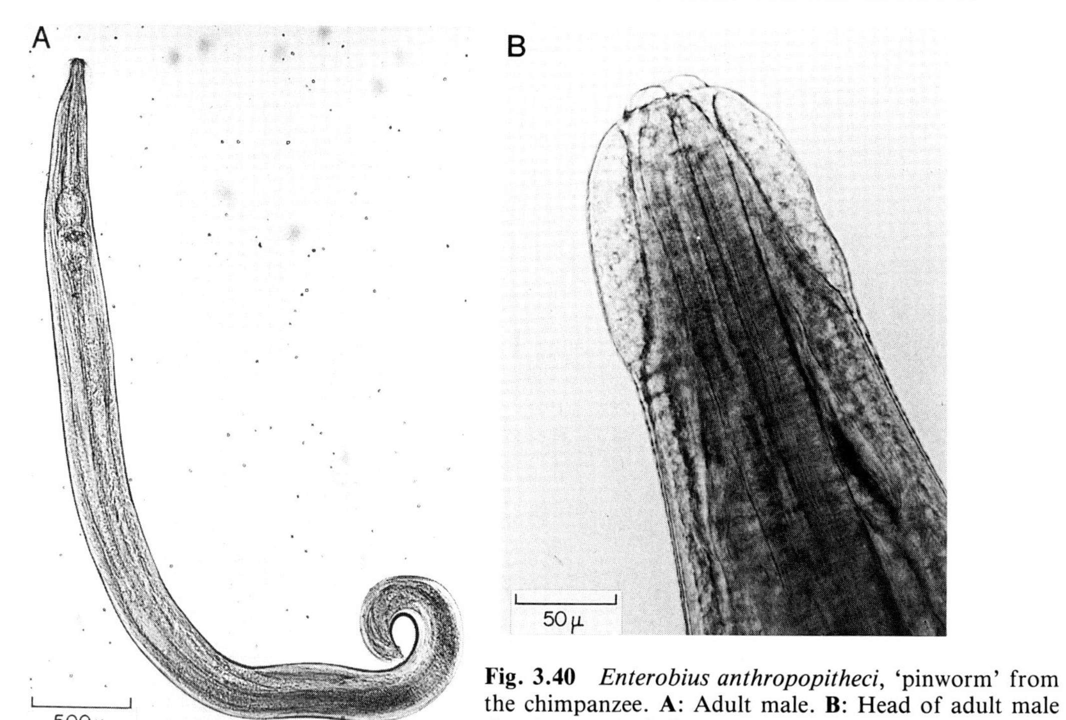

Fig. 3.40 *Enterobius anthropopitheci*, 'pinworm' from the chimpanzee. **A**: Adult male. **B**: Head of adult male showing cervical alae.

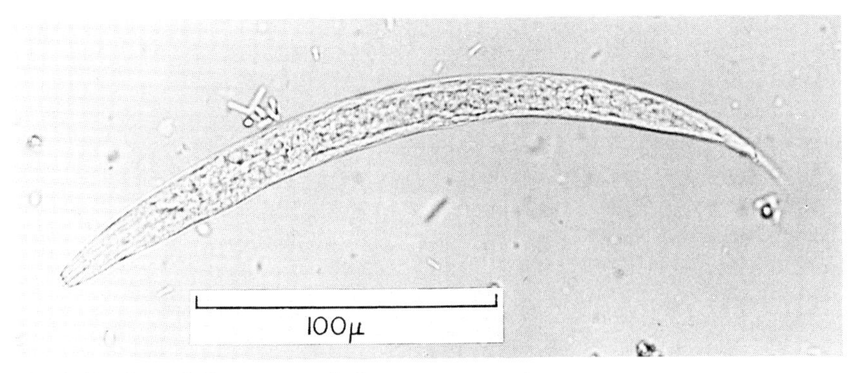

Fig. 3.41 Larval *Trypanoxyuris* from a marmoset.

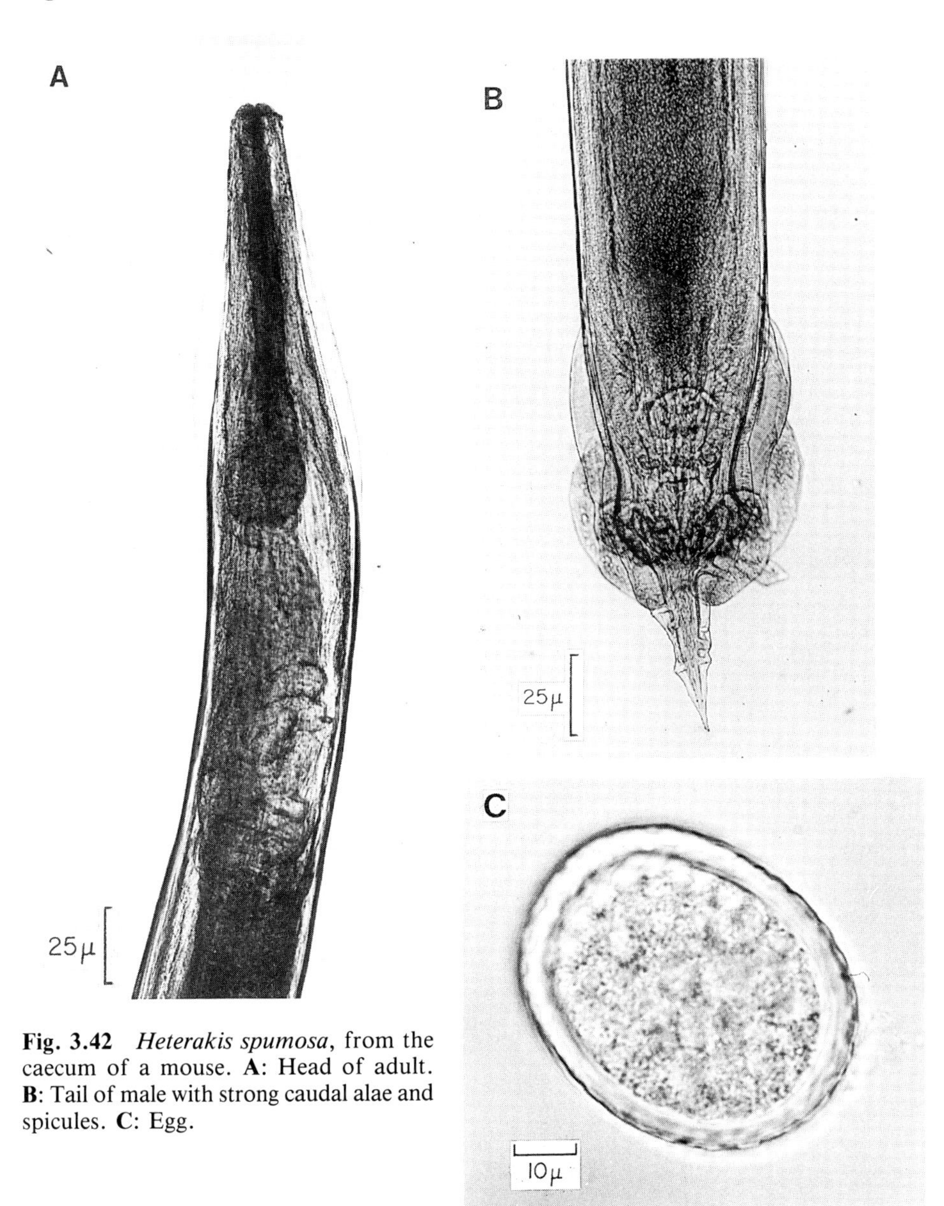

Fig. 3.42 *Heterakis spumosa*, from the caecum of a mouse. **A**: Head of adult. **B**: Tail of male with strong caudal alae and spicules. **C**: Egg.

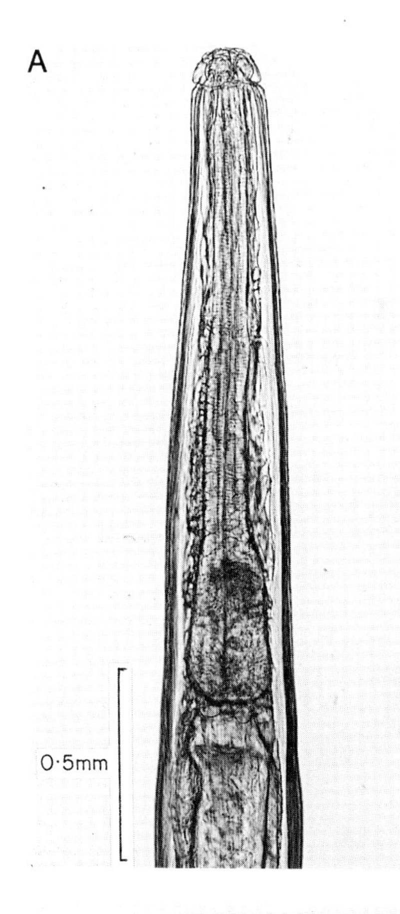

Fig. 3.43 *Paraspidodera uncinata*, from the caecum of a guinea pig. **A**: Head of adult. **B**: Curled tail of male.

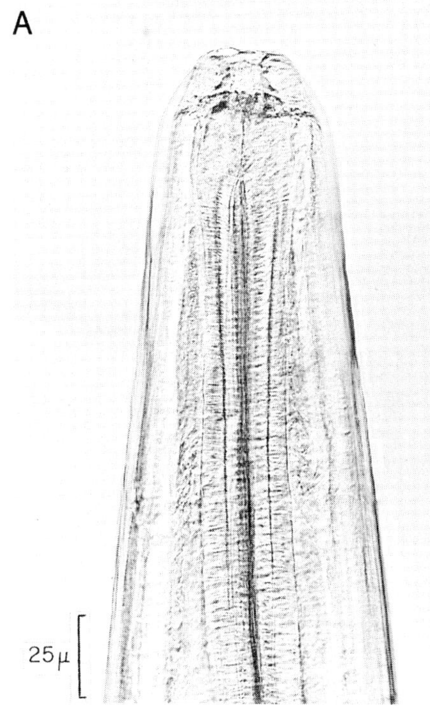

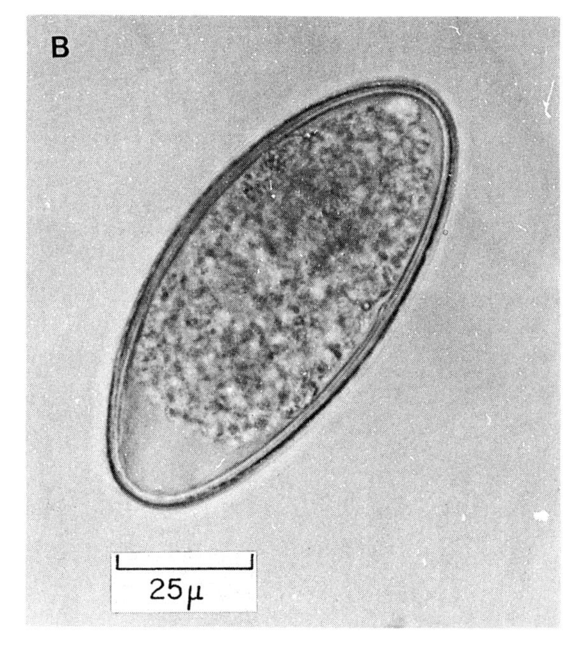

Fig. 3.44 Pinworms from a marmoset. **A**: Head of *Subulura jacchi*. **B**: Egg of *S. jacchi*.

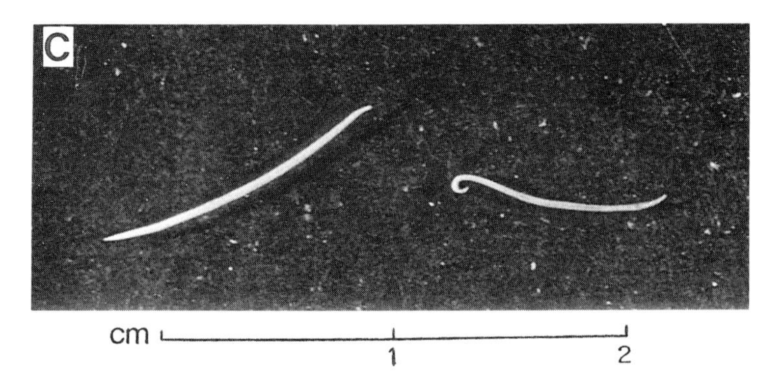

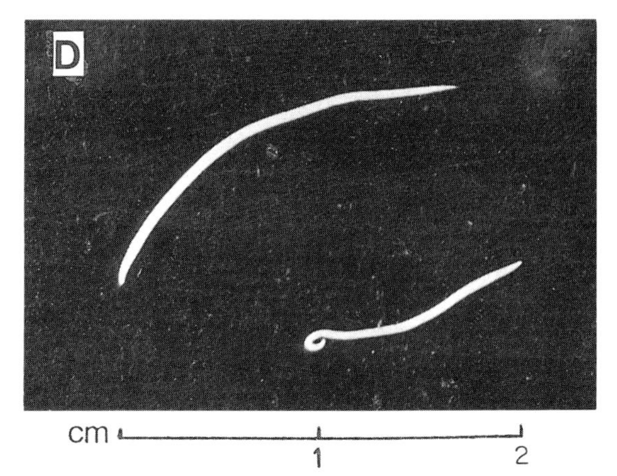

Fig. 3.44 (continued). **C**: *Tarsubulura* sp; adult female and smaller male. **D**: *Parasubulura jacchi*; adult female and smaller male.

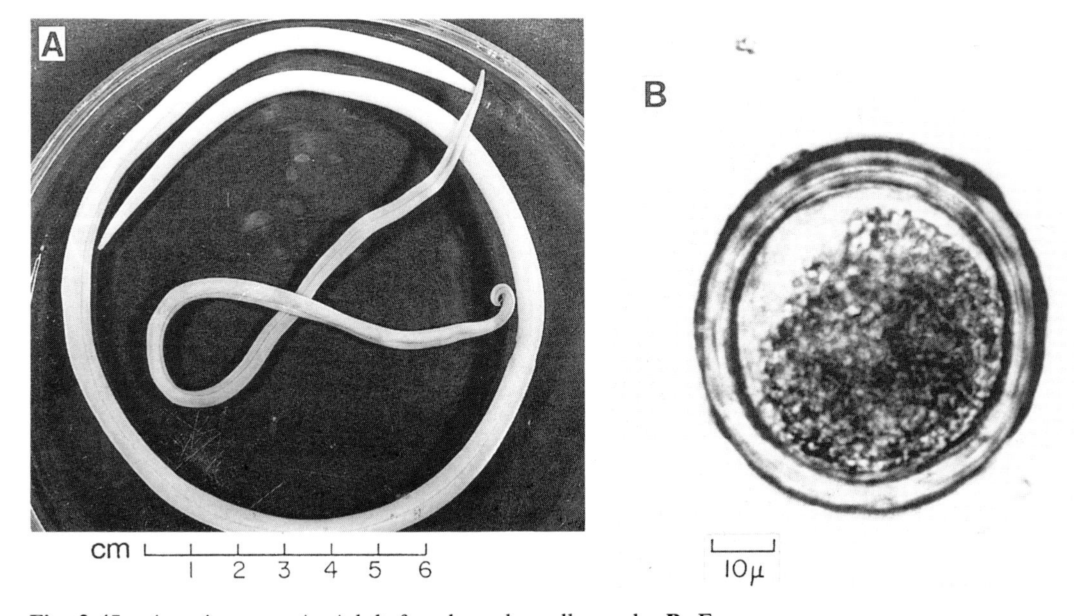

Fig. 3.45 *Ascaris suum*. **A**: Adult female and smaller male. **B**: Egg.

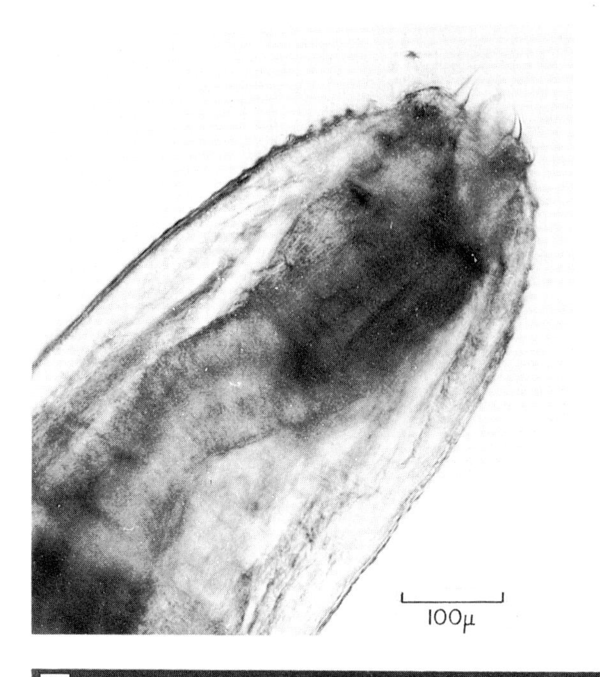

Fig. 3.46 Head of *Protospirura* from *Cebus capucinus*.

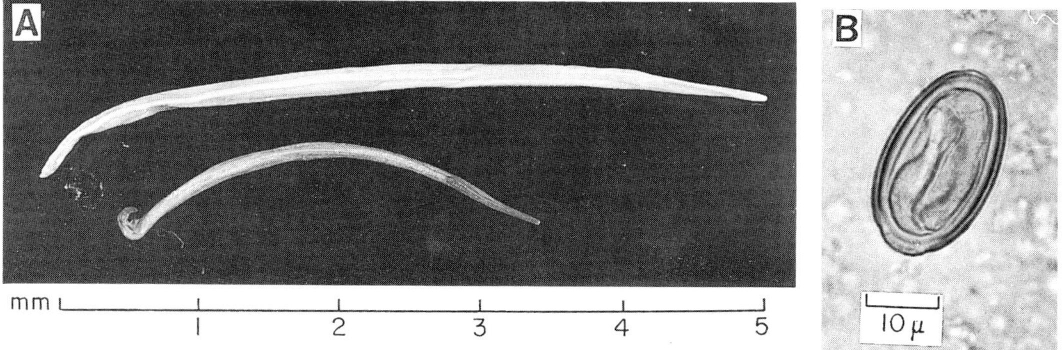

Fig. 3.47 *Physaloptera tumefaciens* from a cynomolgus monkey. **A**: Adult female and smaller male. **B**: Egg.

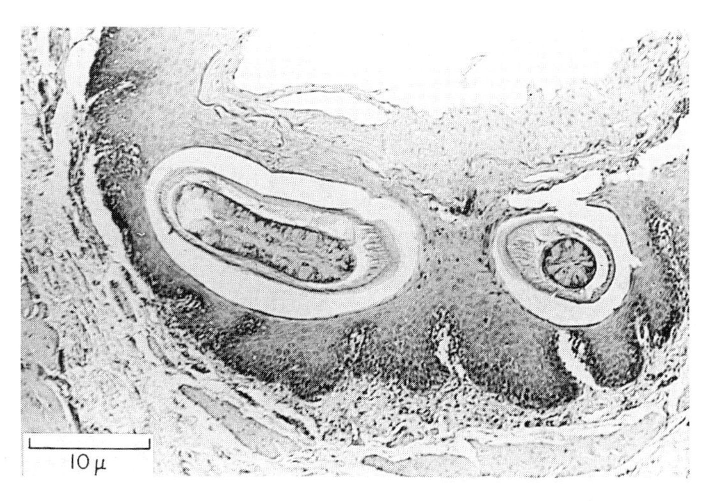

Fig. 3.48 *Gongylonema* sp: adult in the oesophagus of a macaque.

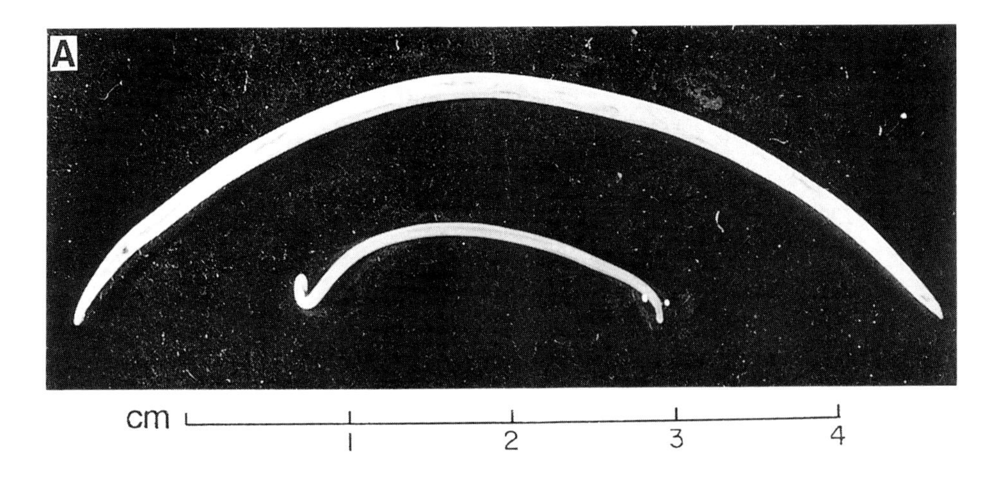

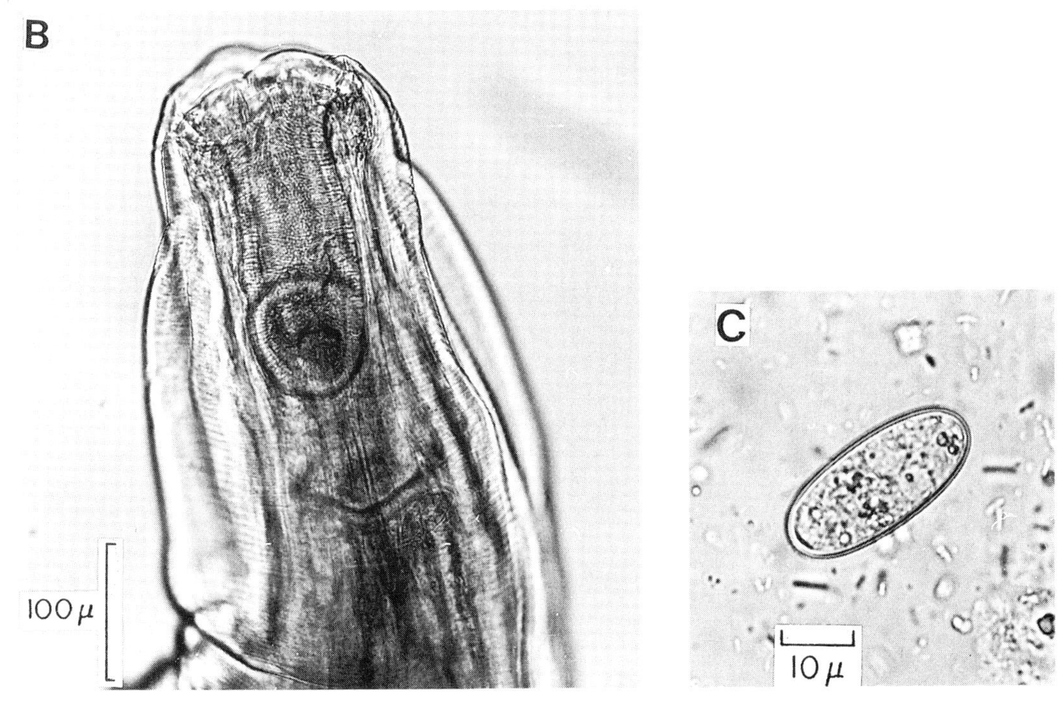

Fig. 3.49 *Streptopharagus* sp. from the stomach of a macaque. **A**: Adult female and smaller male. **B**: Head of adult showing typical twisted vestibule. **C**: Egg.

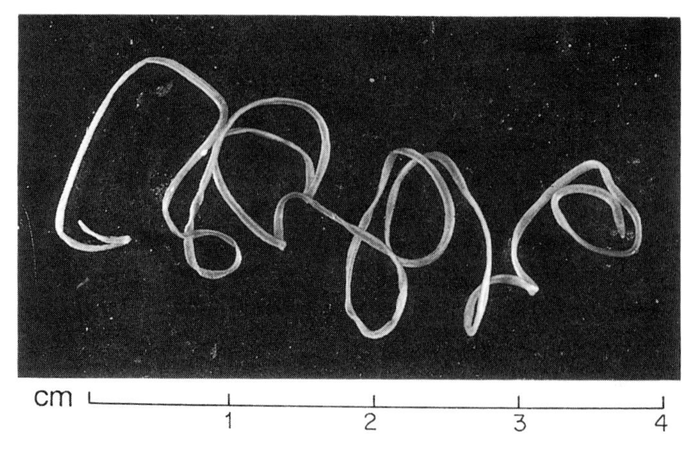

Fig. 3.50 Adult *Dirofilaria* sp. from *Cercopithecus*.

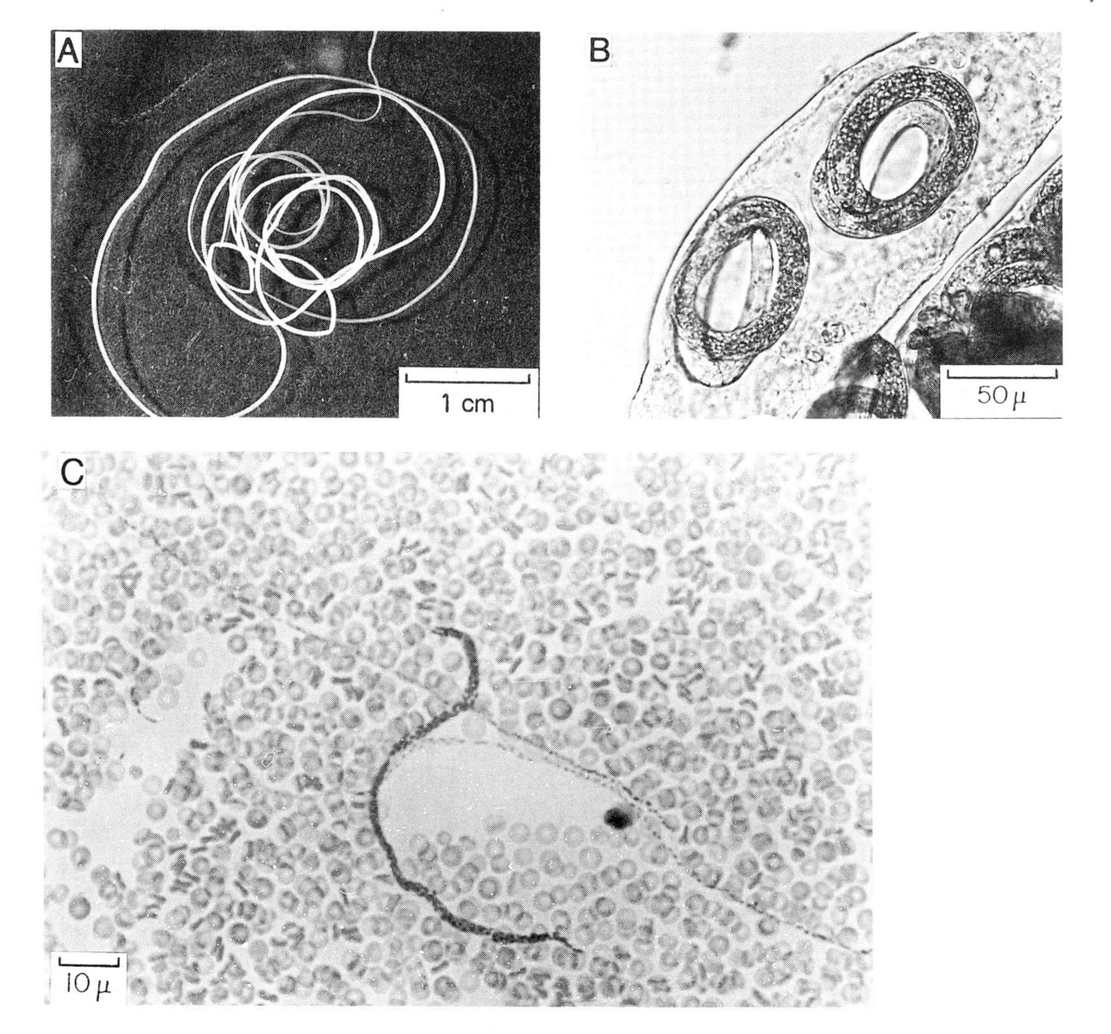

Fig. 3.51 Filarial worms from a squirrel monkey. **A**: *Dipetalonema* from the body cavity. **B**: Microfilariae in the uterus of the female. **C**: Microfilariae of *Tetrapetalonema* (long and fine) and *Dipetalonema* sp.

ORDER STRONGYLORIDA

This particular group of nematodes is characterized by the possession of a cuticular bursa supported by muscular rays. They include some of the most damaging intestinal nematodes of man and domestic animals. In this group are the nodular worms, hookworms, lungworms and many others which are found in a large number of both Old and New World primates, but very few species are found in rodents.

Subfamily Oesophagostominae

The most common and most injurious nematodes of the order Strongylorida are the oesophagostomes found among monkeys and apes. They are fairly small worms, the female measuring around 20 mm and the male 15 mm, usually found in the lumen of the large intestine. Long cervical alae have the effect of producing a dorsal curvature of the anterior region, which also has a collar-like ring of cuticle around the mouth which is separated from the rest of the body by a distinct and very characteristic constriction. The buccal cavity is shallow and the margin edged with cuticular structures called leaf crowns or corona radiata which are used to lacerate the mucosal wall. A second inner crown is present within the buccal cavity. The male bursa is well developed and two equal spicules are present. The eggs passed in the faeces are thin-walled and distinctive and are usually at the 8 or 16 cell stage. Outside the host the egg embryonates and if conditions are suitable reaches the infective L_3 stage at 6–7 days.

Infection is by ingestion of the infective larva, which once inside the host proceeds to shed its sheath and burrows into the mucosal wall. In a primary infection of *Oesophagostomum* sp. (Fig. 3.52, page 111) the larva remains in the mucosal wall where it moults and grows over a period of 5–6 days, and then emerges into the lumen. Later invasions produce the characteristic nodules which are formed by the invading L_3 larvae becoming trapped by the strong immunological response from the host. As long as the nodules remain sterile, there is little pathological consequence. A high level of infection with this parasite results in a large area of the mucosa becoming hardened, and peristalsis and digestion being affected. Larvae trapped in those nodules may remain alive for three months or more before caseation of the contents and their eventual death. The major pathogenic effects of this worm are a direct result of the accumulation of these nodules: not only are the digestive processes affected, but there is a possibility of secondary infection of the nodule, leading to abscess formation and consequent peritonitis should the abscess rupture.

The young worms emerging from the nodules grow to adults in the lumen of the large intestine and the eggs first appear in the faeces approximately 40 days post-infection. Reports of *Oesophagostomum* sp. in New World species are rare, but they are very common in Old World primates including macaques, baboons and the great apes. Many species have been named but *O. apiostomum* appears to be the most common among cynomolgus monkeys and other macaques.

Closely related are species of *Ternidens* (Fig. 3.53, page 112) which are found in the caecum and large and small intestine of a wide variety of guenons and macaques as well as chimpanzees and man, and are fairly common in the wild state, but rare in laboratory situations. Unlike *Oesophagostomum*, it is a blood-sucker and has been described as causing anaemia and cystic nodules in the colon (Levine 1968). The adult worms are very similar morphologically to *Oesophagostomum* (Tanaka *et al.* 1962). Distinguishing the eggs of *Ternidens* from those of *Oesophagostomum* is not easy. Kuntz *et al.* (1968)

distinguished between them in their survey by means of size, separating them into two size groups: *Oesophagostomum* measuring 48–56 × 32–33 μ (mean 52 × 32 μ), and *Ternidens* measuring 73–92 × 44–78 μ (mean 88 × 46 μ). The buccal capsule of *Ternidens* is globose and contains jutting protruberances rather than the teeth which are typical of the hookworm family (Sandground 1931).

Family Ancylostomatidae

Hookworms are not common parasites of monkeys and apes, and the very few reports of their presence in South American monkeys could be due to an error of identification as the eggs of many species of strongyles are morphologically very similar. They have, however, been reported as naturally occurring in anthropoids and have been experimentally transmitted in chimpanzees (Orihel 1972). Both *Ancylostoma duodenale* (Fig. 3.54, page 112) and *Necator americanus* of man have been reported in anthropoids but their presence is rare.

Other species of ancylostomes are common in dogs and cats throughout the world. The life cycle is direct: the L_3 infective larvae are either ingested or bore their way through the skin. On entering the body, they migrate to the lungs, penetrate the alveoli and when they reach the trachea they are eventually swallowed. The adult in the small intestine is small, the female 15–18 mm, the male 9–12 mm (Soulsby 1965), with a distinctive shallow buccal cavity containing well-developed cutting teeth. The egg is ellipsoidal and usually in the 2–8 cell stage when passed, measuring 55–72 × 34–45 μ. The rate of development from first stage to infective larva depends upon the general environmental conditions.

Family Trichostrongylidae

Members of the family Trichostrongylidae are distinguished from other members of the order Strongylorida by their small size and the absence of, or a very small, buccal capsule. They have no teeth or leaf crowns, and are parasites of the alimentary tract. Two common trichostrongyles, *Heligmosomoides* (*Nematospiroides*) *dubiuus* (Fig. 3.55, page 112) and *Nippostrongylus brasiliensis*, are parasites of the wild rat and mouse and are often used as models of human hookworm disease and for the study of immunity in helminthiasis.

The adults of *H. dubiuus* are found in the small intestine of the mouse. They are small, bright red in colour and the females are tightly coiled into a spiral. The male is shorter, measuring 4–6 mm, and is less tightly coiled. Typically the posterior end of the male has a broad copulatory bursa with chitinous supporting rays and two long tapering spicules. The eggs, measuring 70–84 × 37–53 μ, are thin-walled and ellipsoidal in shape; they are usually at the 8 or 16 cell stage of development, are passed with the faeces, and require 26 hours to embryonate and hatch even with suitable environmental conditions. The newly hatched larvae measure 480–503 μ in length; they moult to second-stage larvae 48 hours after hatching and to third-stage infective larvae in 4–6 days. The infective larva retains the cuticle of the second stage as a sheath and remains quiescent until stimulated by vibration or air currents. Infection of the host is by ingestion of this stage. Once in the stomach the larva sheds its sheath and migrates to the small intestine where it can be found burrowing into the mucosa 24–48 hours later. The larva undergoes a fourth moult to the adult stage at 6–8 days; egg-laying starts 9 days post-infection and may continue for up to 8 months. *H. dubiuus* is not a common parasite of laboratory rodents but is found regularly in the wild population. It does not appear to be pathogenic except in large numbers and the effect on the host is thought to be due to a toxin produced by the worm.

Nippostrongylus brasiliensis differs from *H. dubiuus* in a number of ways despite their similar structure and location in the host. The adult female is not coiled and there is little difference in size between the sexes. The female measures 2.5–6.2 mm and the male 2.1–4.5 mm. There is a simple oral cavity and a slight inflation of the cephalic cuticle. The egg is ellipsoidal and measures 52–63 × 28–55 μ. As in *H. dubiuus*, the eggs are laid at the 8–16 cell stage and embryonation and hatching take 24 hours. Development takes place over 3–4 days, but infection of the host is usually by penetration by the larvae through the skin. The larvae next migrate through the tissues, progressing through the lungs and reaching the alimentary tract via the trachea and oesophagus. There is a prepatent period of 6–8 days and the infection can remain patent for several months. Symptoms only occur where there are very heavy infections; there may then be skin irritation from larval migration and inflammation of the lungs and small intestine. If infection is sufficiently heavy, this may result in the death of the animal.

Trichostrongylus sp. (Fig. 3.56, page 113) are common parasites of the small intestine of herbivores and are also found in rodents, primates and man. They are small, delicate worms, the females measuring 5.0–7.0 mm and the males 4.5–5.0 mm. The egg is oval, slightly tapered at one end and is at the 16–32 cell stage when laid; it measures 75–86 × 34–45 μ. The life cycle is simple and direct, with the larva hatching at around 24 hours when environmental conditions are suitable. It reaches the third, infective stage 3–4 days later. Infection of the definitive host is by ingestion of the third-stage larva and development continues as in other species described above.

A closely related species is the stomach worm of the rabbit, *Graphidium strigosum*. This is common in the wild but not usually found in laboratory-bred animals. The female measures 11–20 mm and the male 8–16 mm. The eggs measure 98–106 × 50–58 μ and are not embryonated when passed in the faeces. Embryonation takes 24–36 hours and development to the infective stage 4–6 days. Infection of the host is by ingestion of the infective larvae which develop in the wall of the stomach over a period of 12 days. In both *Trichostrongylus* and *Graphidium* sp. light infections appear to produce little if any clinical effect, whereas heavy infections cause diarrhoea, anaemia and emaciation (Adams *et al.* 1967; Lapage 1968; Faust *et al.* 1968).

There is a range of trichostrongyle species found in wild-caught monkeys. A common parasite of South American monkeys is *Molineus torulosus* (Fig. 3.57, page 113) which has been reported in the capuchin and squirrel monkey in Brazil (Dunn 1961, 1968); and *M. vexillarius* is found in the stomach and small intestine of tamarins in Peru (Cosgrove *et al.* 1968). Many other species have been found but are less well recorded. *Molineus* sp. are small, pinkish worms, the females measuring 3.26–5.3 mm in length and the males 3.0–4.8 mm. The egg is typical, being small and at the 8–16 cell stage measuring 40–52 × 20–29 μ. The life cycle is not known but infection is thought to be by ingestion of the infective third-stage larvae. In large numbers, *M. torulosus* is a pathogenic worm causing extensive haemorrhage and ulcers.

One of the most pathogenic species is *Nochtia nochti*, which infests the stomach of *M. mulatta* (Travassos & Vogelsang 1929), Javanese macaques (Bonne & Sandground 1939) and *M. speciosa* from Thailand. This worm has been associated with extensive gastric tumours and papillomata. They are filariform worms, the female measuring 7.6–9.9 mm and the male 5.7–6.5 mm. These bright red worms can be seen protruding from the surface of the tumours which are found in the prepyloric region of the stomach (Smetana & Orihel 1969).

Family Metastrongylidae

These are long, slender, delicate worms that inhabit various portions of the respiratory system of a range of mammalian species. Those that parasitize the bronchioles of primates belong to the genus *Filaroides*, of which there are four named species (*see* Appendix A3). Typically the adult females in the alveolar ducts and sacs produce live larvae which are coughed up and swallowed by the host. The parasites produce little in the way of inflammation. The gross appearance of the lung is normal except that occasionally nodules may be apparent at post-mortem examination and these may be pigmented. Infection can be diagnosed by finding the hatched first-stage larvae in very fresh stool samples. To date there has been no evidence that infection with *Filaroides* sp. produces any pathogenic effects (Dunn 1968).

ORDER RHABDITORIDA

The most important genus of this order is *Strongyloides* (Fig. 3.58, page 114). These nematodes are an unusual group within an order which is primarily free-living, and have within their life cycle both a parasitic, parthenogenic female stage and a free-living adult male and female stage. They are very small worms, the parthenogenic female measuring 2–5 mm × 30–80 μ. This stage is found in the mucosa of the duodenum and jejunum; it is filariform with a cylindrical oesophagus filling the anterior third of the body and the vulva opens in the posterior third. The eggs are thin-walled and embryonated when laid, measuring 40–70 × 20–35 μ.

The egg hatches rapidly and the larva is rhabditiform in structure, measuring 150–340 μ in length with a distinctive muscular oesophagus constricted before the bulb. The free-living male measures 0.8–1.1 mm in length with a short conical tail, gubernaculum and two short equal spicules but no caudal alae. The female measures 0.9–1.7 mm and has a long pointed tail, and the third stage filariform infective larva measures 400–800 × 12–20 μ. The distinctive feature of the infective stage is the long cylindrical and non-bulbous oesophagus which occupies the complete anterior half of the worm. At the posterior extremity, three small points at the tip of the tail give it a distinctive notched appearance.

There have been many theories as to whether development can be direct, that is the rhabditiform larva becomes filariform without a soil stage; or indirect, in which there are one or more free-living cycles before filariform larvae are produced. They are prolific, however, and highly infectious and when filariform larvae are produced, infection of the host is by penetration of the skin or oral mucosa. Next begins a stage of migration during which the larvae progress via the bloodstream to the heart and lungs where they moult. The next stage progresses via the alveoli and bronchii into the mouth and is swallowed (Faust 1933). The last moult occurs in the duodenum and the worm develops into an adult parthenogenetic female.

S. fulleborni is a common species found in macaques, baboons and chimpanzees in Africa and Asia (Ruch 1959; Rowland & Vandenbergh 1965; Van Riper *et al.* 1966; Reardon & Rininger 1968); and *S. cebus* is found in a range of New World species in Central and South America (Little 1966). *S. stercoralis* is the species found in man, the chimpanzee and other great apes (Desportes 1945). *S. ratti* has been described in the Norway rat (Ash 1962; Dove 1950; de Léon 1964) but has not been reported in barrier-maintained colonies (Foster 1963).

In all cases low infections result in few, if any, clinical signs, whereas heavy infections result in skin irritation at the site of penetration, a dry cough at the stage of lung invasion and severe gastric disturbance as the worms penetrate the gut wall.

ORDER DORYLAIMORIDA

Family Trichuridae

This group differs from other nematodes described so far in posessing a non-muscular oesophagus, made up of glandular stichocyte cells, and having no excretory canals.

Adult *Trichuris* species are characterized by marked structural differences between the long slender oesophageal portion which takes up three-quarters of the body length, and the short thick posterior region which contains the reproductive organs. They are known as the 'whipworms' because of this odd morphological feature (Fig. 3.59, page 114). The slender 'neck' portion is usually found embedded in the mucosa, so the worms are not easily dislodged from the gut wall. *T. trichiura*, the human whipworm, has been found in a range of Old World primates, and the species *T. muris* has been isolated from wild rats and mice.

There is little difference in size between the sexes, both measuring 45–75 mm in length. The male has a distinctive, tightly coiled posterior tip with a single spicule enclosed in a spiny terminal sheath, and the vulva of the female is situated at the junction of anterior and posterior portions.

The life cycle is simple and direct. The dark brown oval eggs are distinctive with their thickened barrel-shaped walls plugged at each end. They measure 70–80 × 32–40 μ and are not embryonated when passed in the faeces. The female is a prolific egg producer—as many as 2000 in one day (Miller 1947). Development outside the host is dependent on environmental conditions, taking between 9 days and 4 weeks. The larva remains within the egg and hatches when swallowed; with the aid of a small retractable lancet it then penetrates the mucosa of the small intestine where it continues to develop and emerges back into the lumen 2–10 days later. From there it passes to the caecum and colon and the adult worm is mature at 3–4 months. The longevity of this species is thought to be 12–18 months.

As with most enteric nematodes, light infections of *Trichuris* produce little or no clinical signs, whereas heavy infections produce a pale mucoid diarrhoea and prolapsed rectum, and can result in death (Graham 1960; Thienpont *et al.* 1962).

Capillaria hepatica is a closely related species common in the wild rat, with a few reported infections in laboratory rodents (Ash 1962; Fisher 1963). It has been found in a wide range of mammalian hosts including man. Primates are not common hosts but it has been found in rhesus monkeys, capuchins and apes (Ruch 1959) where it can be pathogenic (Fiennes 1967). They are long, slender worms which are found threaded through the liver parenchyma. The females measure up to 100 mm in length and 200 μ in width and the males 17–22 mm × 40–80 μ (Olsen 1967). There is a simple mouth and the oesophagus has a long glandular portion typical of this group. The male lacks the single spicule but a protrusable sheath is evident, and the prominent vulva of the female lies in the oesophageal region. The lemon-shaped plugged eggs are extruded into the liver tissue and the fibrous tissue reaction around them can be clearly seen in post-mortem examination as patches and nodules of white tissue distorting and scarring the liver surface (Fig. 3.60, page 115).

The longevity of the worm is short once it reaches its host, but the life cycle is a tortuous one. The eggs extruded into the liver tissue remain viable for a considerable time and are only released when the host dies, reaching the outside world either by passage through a predator or by decomposition of the host's body. Under suitable aerobic conditions they require 4–6 weeks to embryonate. The final host is infected by ingestion of the embryonated egg and the larva hatches in the upper small intestine and proceeds to penetrate the mucosal wall. The larvae reach the liver over a period of 2–10 days and develop to mature adults in 30 days. The adults live only a few weeks following maturity, and thus sections of the fibrosed surface of the liver often show only masses of ova and tissue damage. The patchy masses of eggs in the liver cause a certain amount of liver damage, though there are no reports of clinical signs in rodents; in primates, however, they have been associated with a fatal hepatitis.

Trichinella spiralis (Fig. 3.61, page 115) is a cosmopolitan parasite, the infectious stage of which is found in the muscles of mammals. It is not a natural parasite of primates but they are very susceptible to infection. A recent reclassification has restricted *T. spiralis* to the forms found to be zoonotic in man, swine, rats, cats and dogs. *T. nativa* and *T. nelsoni* are the names given to the sylvatic forms of the parasite although they could interact one with the other. The worms are very small, the female measuring up to 4 mm in length and the male 1.5 mm. The female is viviparous and from her position deep in the intestinal mucosa the larvae are liberated into the bloodstream where they are carried to the liver, lungs, diaphragm, eyelid, neck and tongue where they encyst within the muscle fibres. The next stage of the life cycle depends upon the muscle being eaten by another suitable host. Once ingested the larvae are freed into the duodenum, migrate into the mucosa and within a week develop into mature worms which have only a few weeks to live but produce very large numbers of larvae.

Trichosomoides crassicauda, like *Capillaria*, is a very fine thread-like worm found in the wall of the urinary bladder and kidney pelvis of wild and laboratory rats (Fig. 3.62, page 116). The female measures 10 mm in length by 200 μ in diameter; the males are only 1.5–3.5 mm in length and 23–33 μ in diameter and are found inhabiting the genital tract of the female worm. Several males may be found at one time within a single female. The eggs are thick-walled, dark in colour and plugged at each end and measure 60 × 30 μ. They are not embryonated when laid and pass out of the host in the urine. Infection is by ingestion of the embryonated eggs; the larvae hatch in the stomach, penetrate the stomach wall and are carried to the lungs and body cavities within hours. There is a high mortality among the larvae on their migration to the bladder where the fully grown worms can be found in the walls 3–6 weeks later. The complete cycle from ingestion of egg to adult takes a total of 8–9 weeks. These worms appear to have little pathogenic effect on the host, although heavy infections may cause granulomatous lesions in the lungs and they have been thought to cause bladder calculi or to be instrumental in initiating bladder tumours. They can be controlled by strict hygienic measures which prevent development of the ova (Weisbroth & Scher 1969).

Anatrichosoma cutaneum invades the nasal mucosa of rhesus monkeys and has also been reported from man in Asia (Faust *et al.* 1968; Karr *et al.* 1979). It is a very common parasite of wild rhesus monkeys and is best diagnosed by nasal swabbing for the very distinctive eggs which are not always easy to find in the faeces. The eggs are ellipsoidal and typically plugged, measured 60–67 × 40–48 μ and are always embryonated when laid (Fig. 3.63, page 116). Little is known about the morphology or life cycle of the worm but the latter is likely to be direct, the embryonated eggs being sneezed out with mucous secretions. They cause little apparent tissue damage despite the fact that they are often present in large numbers.

A cutaneous form of *Anatrichosoma* was reported by Swift *et al.* (1922) to cause a creeping eruption in the palms and soles of rhesus monkeys. It has also been described in lemurs, marmosets, langurs, macaques, siamangs and orang-utans (Dalgard 1980).

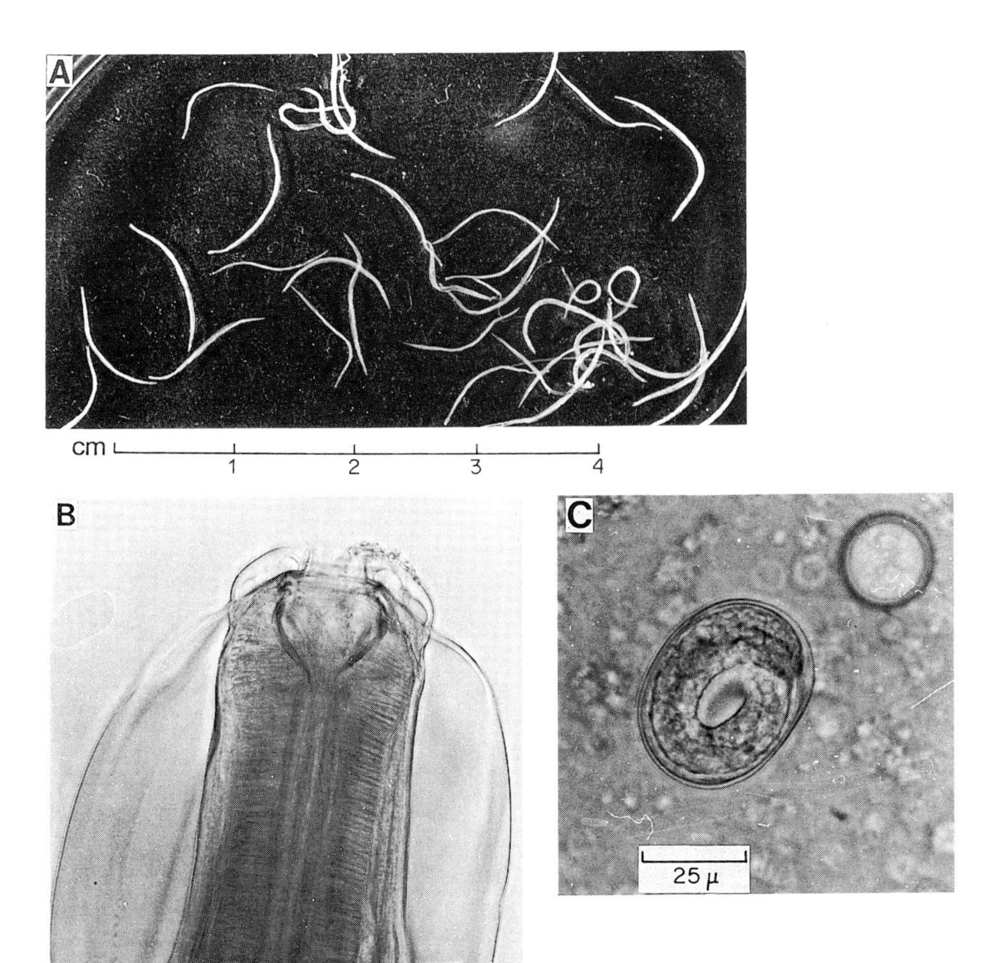

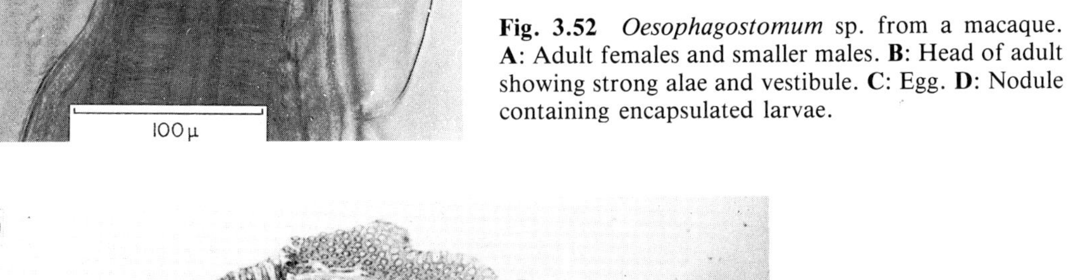

Fig. 3.52 *Oesophagostomum* sp. from a macaque. **A**: Adult females and smaller males. **B**: Head of adult showing strong alae and vestibule. **C**: Egg. **D**: Nodule containing encapsulated larvae.

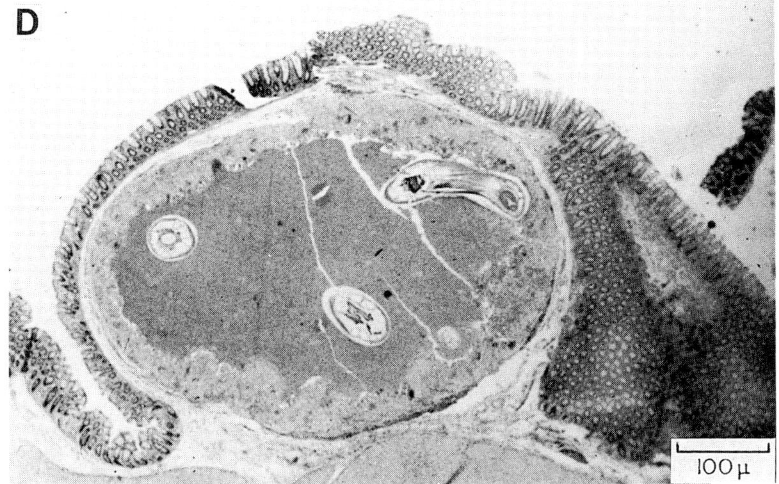

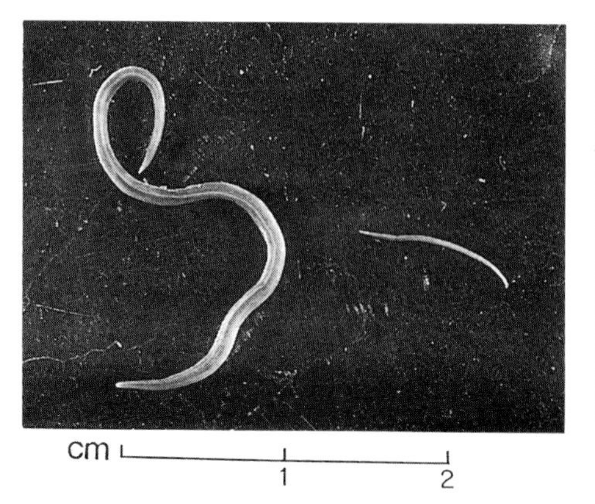

Fig. 3.53 Adult female and male *Ternidens* from a chimpanzee.

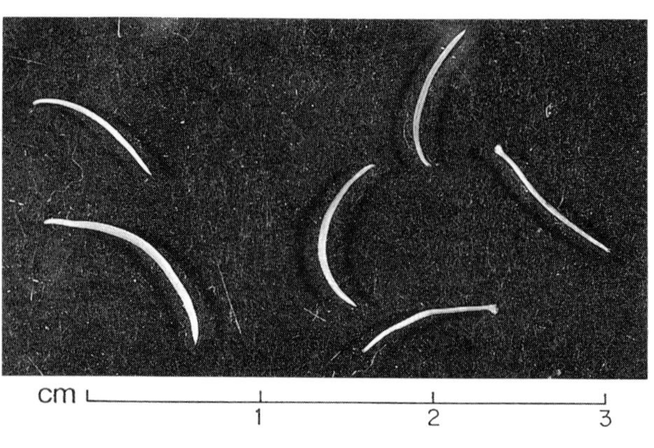

Fig. 3.54 *Ancylostoma duodenale*. Adult females and males.

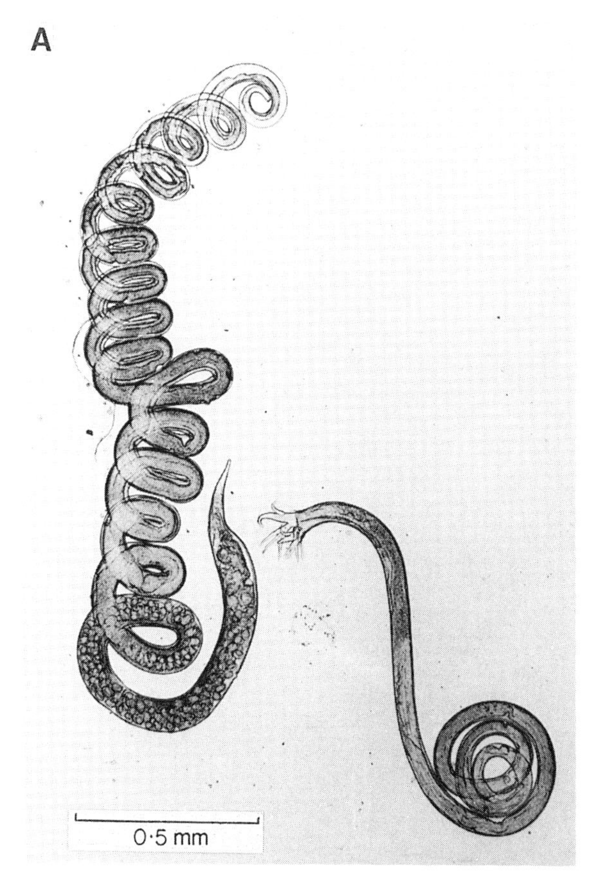

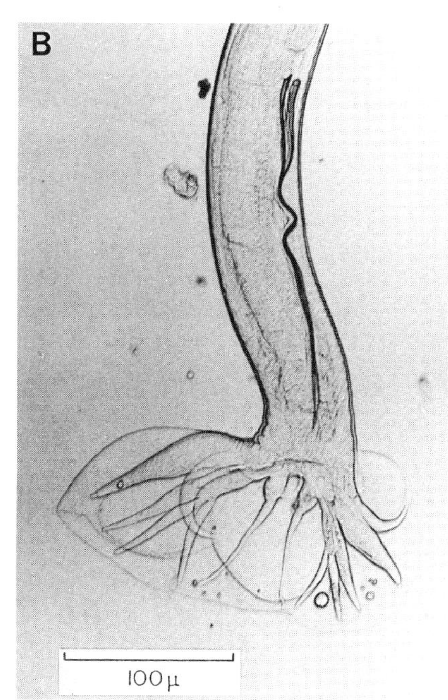

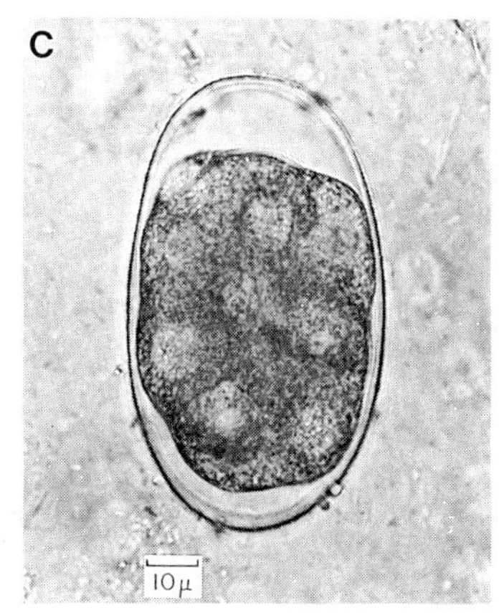

Fig. 3.55 *Heligmosomoides (Nematospiroides) dubiuus* from the small intestine of a mouse. **A**: Adult female and male. **B**: Tail of male showing strong bursa and tapering spicules. **C**: Egg.

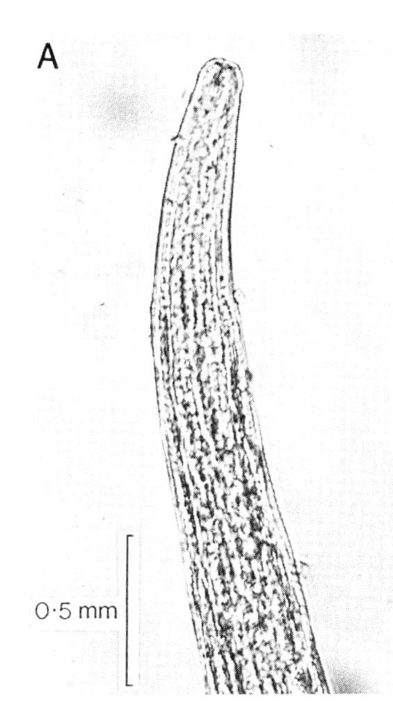

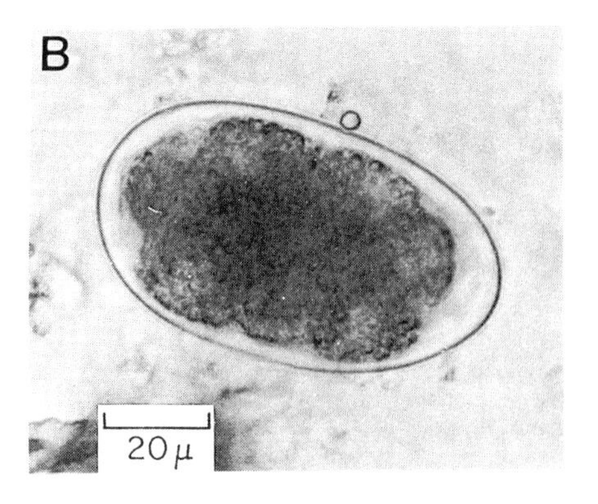

Fig. 3.56 *Trichostrongylus* sp. **A**: Head of adult worm. **B**: Egg.

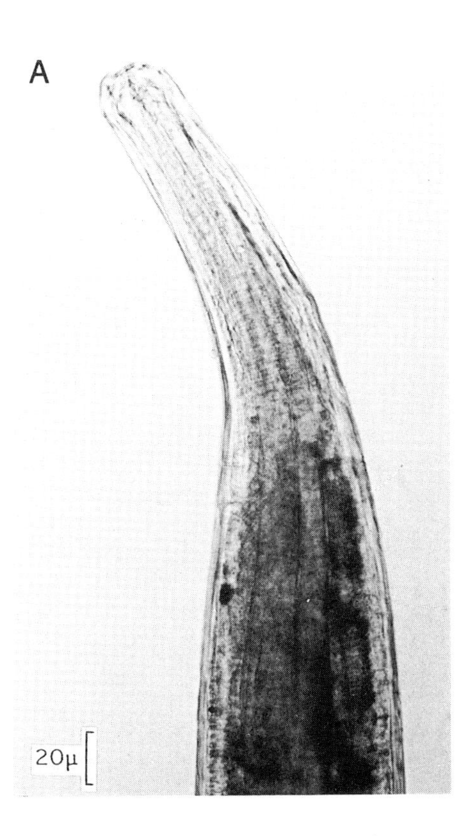

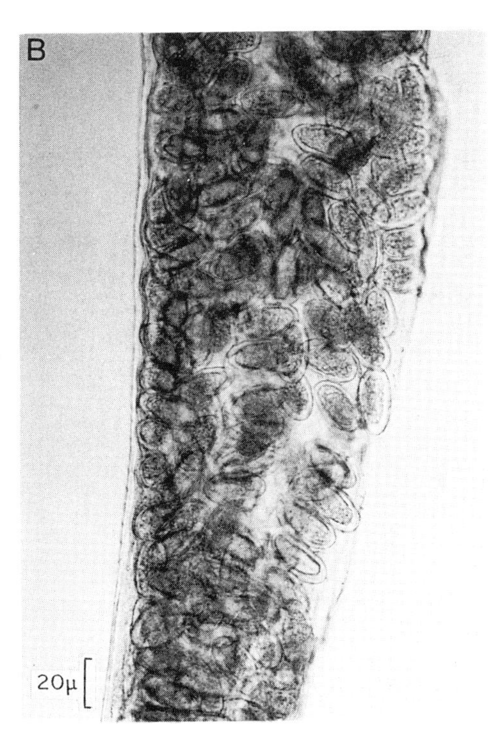

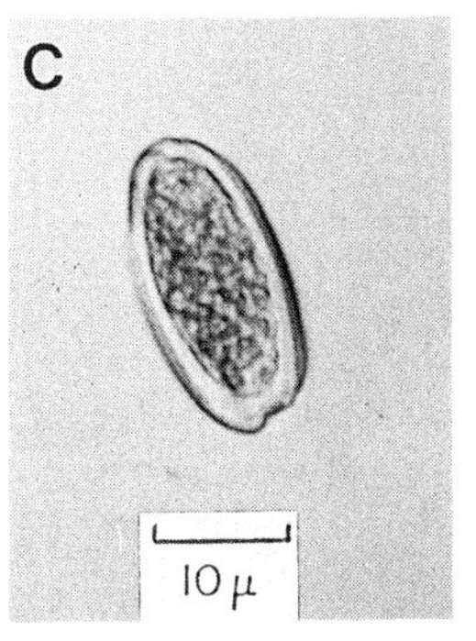

Fig. 3.57 *Molineus torulosus* from a capuchin. **A**: Head of adult. **B**: Female packed with eggs. **C**: Egg.

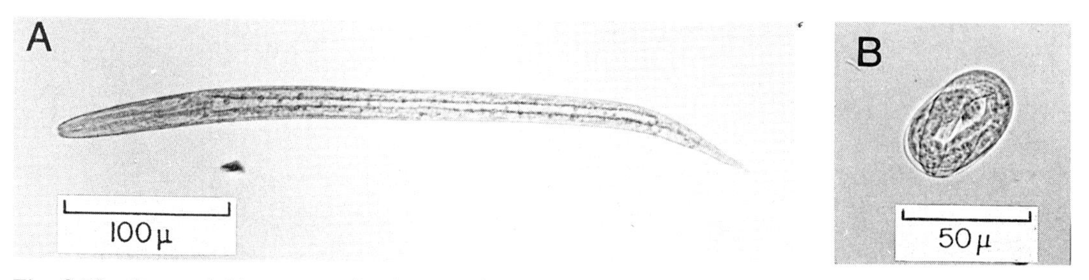

Fig. 3.58 *Strongyloides stercoralis*. **A**: Free-living stage female. **B**: Egg.

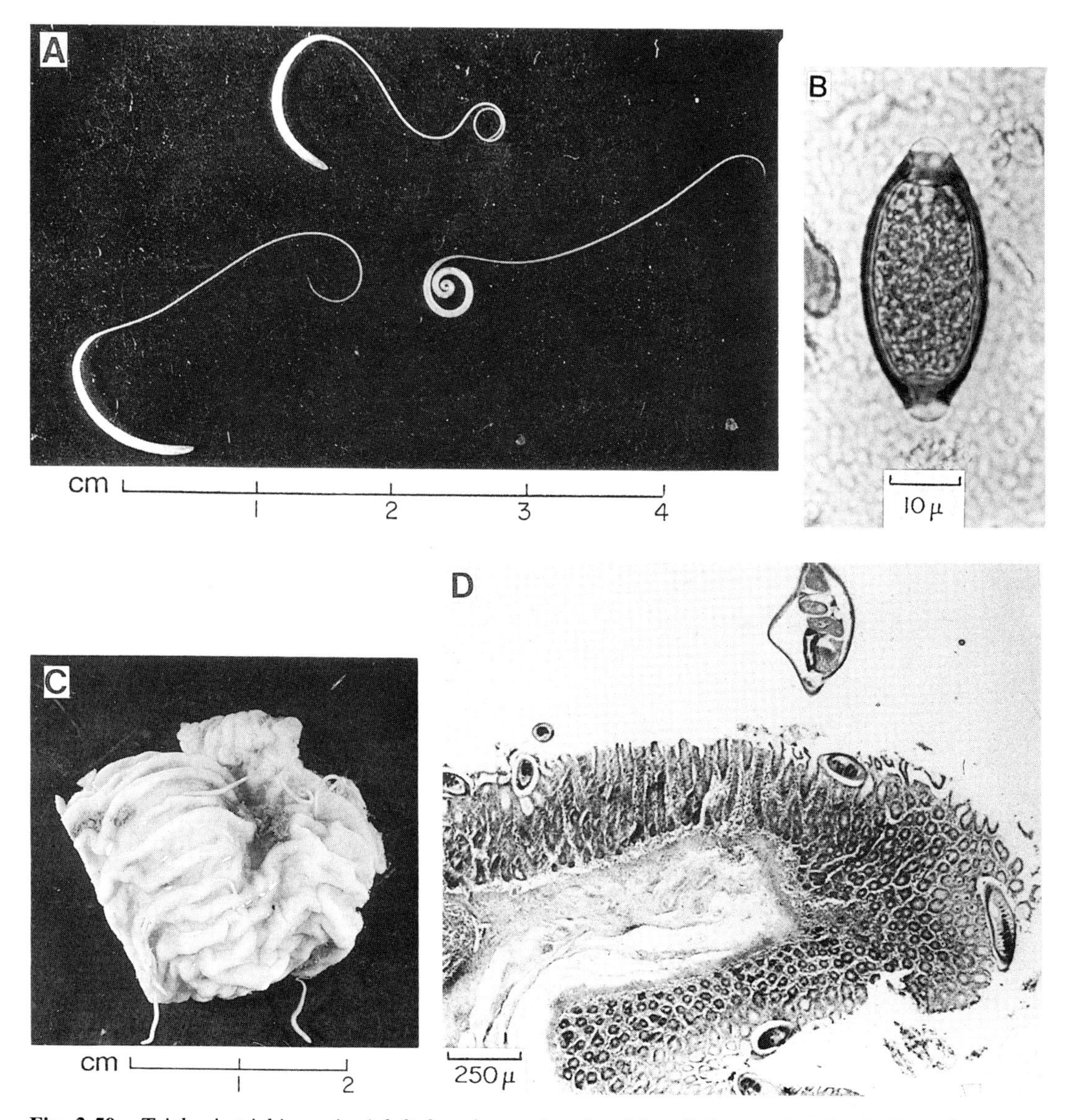

Fig. 3.59 *Trichuris trichiura*. **A**: Adult females, and male with coiled posterior tip. **B**: Egg. **C**: Worms attached to caecal wall of a macaque. **D**: Section through caecal wall showing worms firmly threaded into the wall.

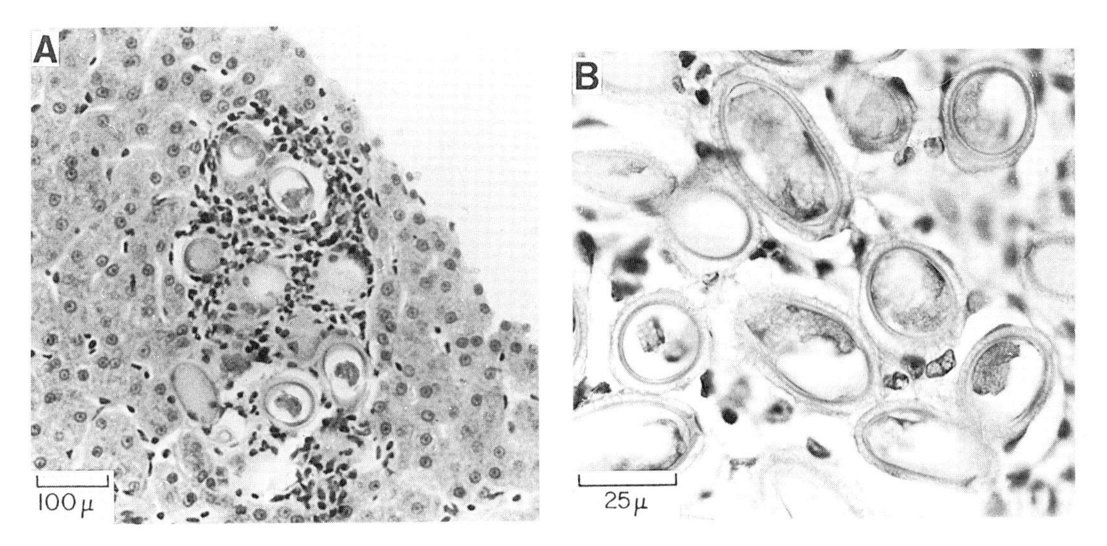

Fig. 3.60 A & B: *Capillaria hepatica* eggs in the liver of a rat.

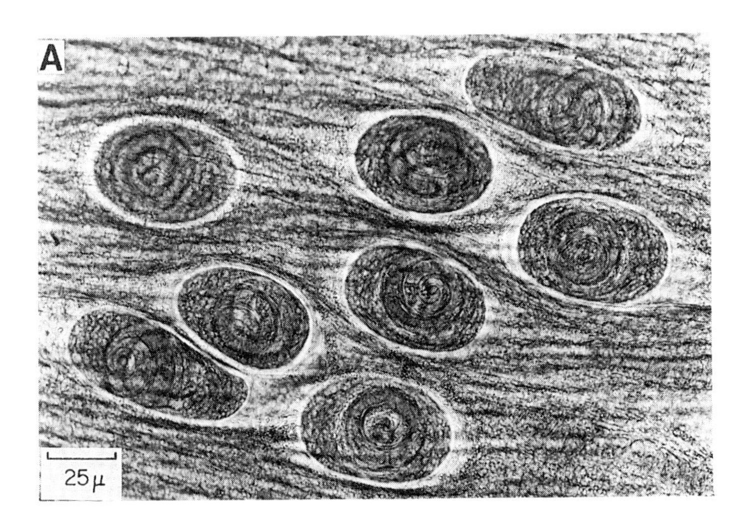

Fig. 3.61 *Trichinella spiralis*. **A**: In the muscle of a rat. **B**: Larvae digested free of muscle.

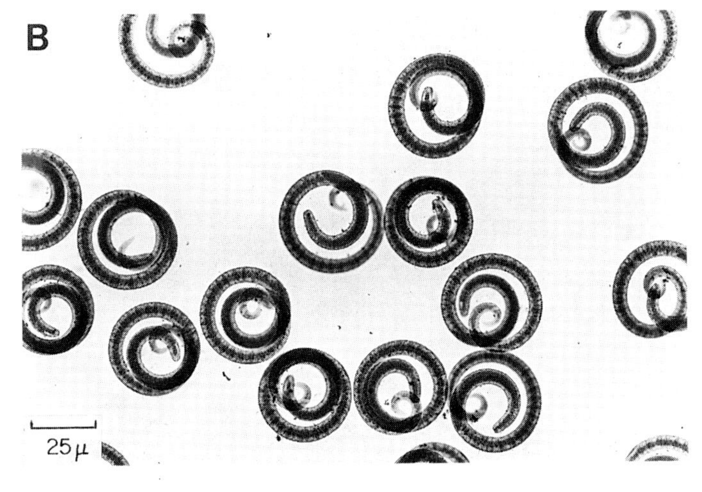

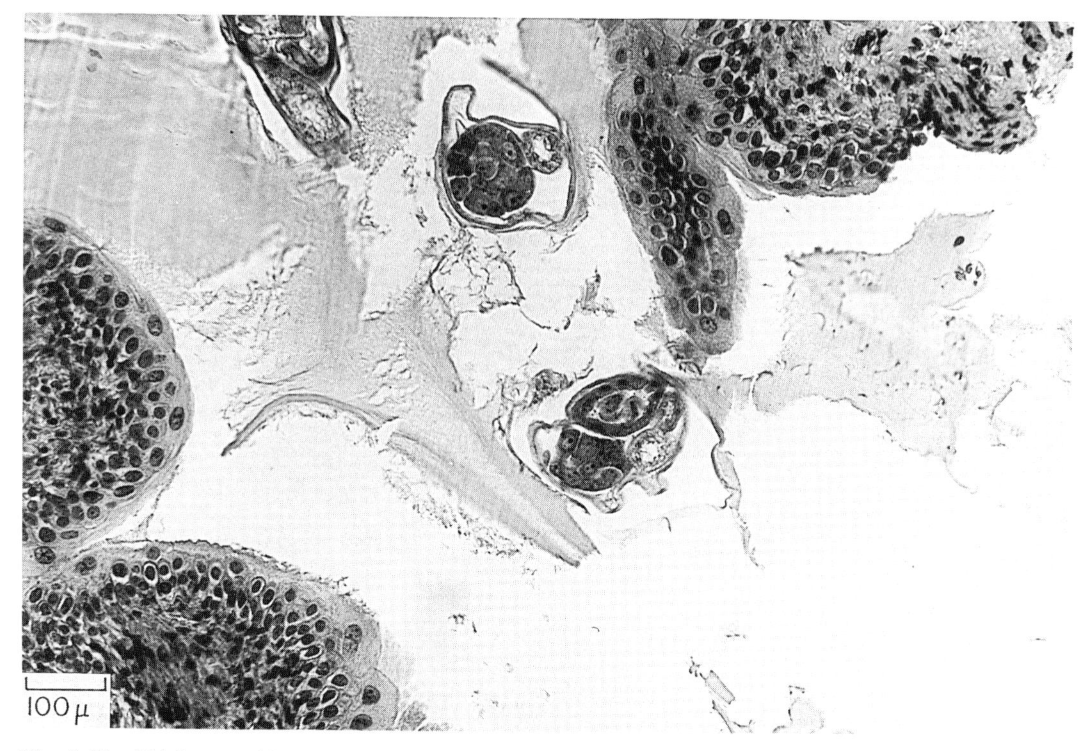

Fig. 3.62 *Trichosomoides crassicauda* in the bladder of a rat.

Fig. 3.63 Egg of *Anatrichosoma* from the nasal mucosa of a macaque.

SECTION 4

TREATMENT

Table 1 constitutes a list of commercially available drugs for the treatment of some of the common ecto- and endoparasites of rodents and lagomorphs together with the recommended dosage. All have been collected by the author from the literature or have been personally recommended as efficacious. It is not by any means a comprehensive list and it should in no way be seen as replacing a need for veterinary advice. It is an attempt to round off the discussion of laboratory animal parasites with some control methods for rodents and lagomorphs which may have been found useful by others and which may fit the circumstances of the reader.

Ectoparasites

Treatment of ectoparasites by dusting or dipping is fairly straightforward but the following points may be of help in ensuring the maximum efficiency of the drug with minimum risk to the health of the animal:

(1) Make up the solution to the manufacturer's instructions and have an adequate size bowl or trough to accommodate the number of animals to be treated.
(2) Make sure the temperature of the solution is constant at 37 °C.
(3) If dipping singly, hold the rat or mouse by the tail and allow it to swim across the bowl dipping the head under once or twice.
(4) If dipping in small groups, place the mice or rats in a wire cage which can be lowered into the tank containing the solution, and submerge the animals as before. The number of animals dipped at one time will obviously depend on the size of container, but it is essential that they should not be overcrowded.
(5) After dipping, place the mice or rats in a clean cage with absorbent material—paper towelling or cellulose wadding—into which they can burrow, and maintain the room temperature at around 24 °C. Drying will take around 1–2 hours.
(6) To prevent recontamination, the dipped animals should be housed in a prefumigated room well away from infested stock.

When contemplating the dipping of a breeding colony, special care should be exercised with young, very old or pregnant animals.

Endoparasites

In the treatment of endoparasites the size of the unit is the usual factor regulating the method of drug administration. Oral dosing of individual mice in a large unit is not practicable, so the incorporation of the drug into the diet could be considered. This method was found to be very successful in the treatment of *Syphacia obvelata* by Thibenzole in an SPF unit (Owen & Turton 1979). More than 100 inbred strains of mice were fed

Table 1. Some commercially available drug treatments for parasitic infection

Parasite	Drug/dose	Mode of application	Manufacturer*
ECTOPARASITES			
ORDER SIPHONAPTERA (FLEAS)			
All species	Tetmosol (monosulfiram BP 25.0% w/w) 1.4%	Dip	ICI
	Note: Sterilization or insecticide treatment of bedding is required to eradicate larval stages		
ORDER PHTHIRAPTERA (LICE)			
Anopleura	Alugan powder (bromocyclen) 0.07%	Dust	Hoechst Animal Health
Amblycera	Alugan bath (bromocyclen) 0.5–0.6%	Dip	Hoechst Animal Health
ORDER ACARINA (MITES)			
Mange-producing fur mites			
Myobia sp.	Alugan powder (bromocyclen) 0.07%	Dust	Hoechst Animal Health
Myocoptes sp.	Lorrexane (γBHC) 6.4%	Dust	ICI
Cheyletiella sp.	Alugan bath (bromocyclen) 0.5–0.6%	Dip	Hoechst Animal Health
	Tetmosol (monosulfiram BP 25.0% w/w) 1.4%	Dip	ICI
Mange-producing follicular and burrowing mites			
Demodex sp.	Alugan bath (bromocyclen) 0.5–0.6%	Dip	Hoechst Animal Health
Sarcoptes sp.	Skin dressing No. 1 (sulphur) 22%	Lotion	Wellcome
Trixacarus sp.	Skin dressing No. 3 (γBHC) 0.625%, two applications	Dip	Wellcome
Psorergates sp.	Lorrexane (γBHC) 6.4%	Cream	ICI
	Tetmosol (monosulfiram BP 25.0% w/w) 5%	Soap	ICI
Otodectic mites			
Psoroptes sp.	PB dressing (Coopers) (piperonyl butoxide 5%, butyl aminobenzoate 2%)	Lotion	Wellcome
	Skin dressing No. 3 (γBHC) 0.625%	Dip	Wellcome
ENDOPARASITES			
PHYLUM PROTOZOA			
Intestinal flagellates (rodents)	Emtryl (dimetridazole) 0.05%–0.1% for 14 days	Drinking water	RMB Animal Health
Intestinal sporozoa (rabbits)	Bifuran (nitrofurazone) 220 mg/litre	Drinking water	Eaton Laboratories
	Pancoxin (amprolium hydrochloride) 20%	Diet	Merck, Sharp & Dohme
	Sulphamezathene (33% solution) 0.2% (30 ml/4.5 l). Course of 5 days. Rabbits can tolerate this dose for several months	Drinking water	Coopers Pitman Moore
	Amprol plus (ethopabate 0.49%, amprolium hydrochloride 7.68%)	Diet	Merck, Sharp & Dohme
CLASS CESTODA			
Intestinal cestodes of rodents	Scolaban (bunamidine hydrochloride) 200 mg/kg single dose	Oral	Wellcome
	Mansonil (niclosamide) 200 mg/kg single dose	Oral	Bayer
PHYLUM NEMATODA			
Intestinal nematodes of rodents	Piperazine citrate 200 mg/kg, 2×7 days with 7 day gap between	Drinking water	Sigma
	Thibenzole (thiabendazole) 0.1%–0.3% for 6–8 days	Diet	Merck, Sharp & Dohme
	Loxon (haloxon) 300 mg/kg single dose	Oral	Wellcome
	Ivermectin 300 mg/kg single dose	Drinking water	Merck, Sharp & Dohme
	Nilverm (levamisole) 20 mg/kg single dose	Oral	ICI

**Manufacturers' addresses:* Bayer, Edmonds, Suffolk, UK; Coopers Pitman Moore, Crewe, Cheshire, UK; Eaton Laboratories, Woking, Surrey, UK; Hoechst Animal Health, Milton Keynes, Bucks, UK; ICI, Macclesfield, Cheshire, UK; Merck, Sharp & Dohme, Hoddesdon, Herts, UK; RMB Animal Health Ltd, Dagenham, Essex, UK; Sigma, Gillingham, Dorset, UK; Wellcome, Berkhampstead, Herts, UK.

pellets of standard rat and mouse diet containing 0.1% Thibenzole for a period of two months. There were no apparent side effects from the drug administration and no differences in breeding performance. Twelve months later there had been no reinfection, indicating a total elimination of the worm.

It is important to note some of the factors that influenced this good result. Firstly, the egg of this worm has a short survival time in the atmosphere, living at the most 2–3 weeks after being laid in optimum conditions. The period of two months allowed for treatment ensured that all the eggs likely to be found in the environment were dead. Secondly, the building was a very strict barrier unit and as the source of original infection had been located and dealt with, the chance of avoiding reinfection was good. In contrast, similar treatment of a conventional or minimal barrier unit may result in only a temporary respite from infection or reduction in numbers.

Where the parasite has a tough, resistant cyst capable of surviving long periods outside the host, eradication is more difficult. Emtryl in the drinking water will reduce the numbers of *Spironucleus (Hexamita) muris* or *Giardia muris*; in the case of mouse strains very sensitive to infection this is sufficient to maintain them free of clinical disease, but only rarely can these flagellates be eradicated. Coccidia are famous for their longevity, and the author has kept samples of *Eimeria* sp. suspended in 2.5% potassium dichromate in the refrigerator for several years and still been able to hatch a good proportion of them. They are not easily destroyed by the ordinary detergents in daily use and only 10% ammonia will certainly destroy them.

Any large-scale treatment is expensive and should be tailored to the facilities and particular requirements of the organization. In all cases the consequences of chemical treatment must be balanced against the pathogenicity of the parasite, and the final decision must be based on the particular priorities of the establishment. Regular and frequent screening is recommended, particularly in high-grade colonies, most of which now originate from caesarean-derived germfree stock. The presence of parasites is readily and easily established and, as they rarely appear alone, is an excellent marker for the possible breakdown of a barrier system with a potentially more serious pathogen.

APPENDIX A: HOST-PARASITE LISTS

Appendix A1. Some ecto- and endoparasites of rodents. (1) = Primary host; (2) = Secondary host

Species	Host	Location	Pathogenicity
ECTOPARASITES			
ORDER SIPHONAPTERA (FLEAS)			
Nosopsyllus fasciatus	Rat	Pelage	None apparent. Vector of *Trypanosoma lewisi* & *T. duttoni*; intermediate host for *Hymenolepis diminuta*, *H. nana* & *H. microstoma*
Nosopsyllus londiniensis	Rat	Pelage	None apparent
Xenopsylla cheopis	Rat	Pelage	None apparent. Intermediate host of plague bacillus *Yersinia pestis*
Leptopsylla segnis	Mouse	Pelage	None apparent. Intermediate host for *H. diminuta* & *H. nana*
ORDER PHTHIRAPTERA (LICE)			
Suborder Anopleura			
Polyplax spinulosa	Rat	Skin surface, blood feeder	None apparent. Vector of *Haemobartonella muris*
Polyplax serrata	Mouse	Skin surface, blood feeder	None apparent. Vector of *Eperythrozoon coccoides*
Suborder Amblycera			
Gliricola porcelli	Guinea pig	General body surface	None
Gyropus ovalis	Guinea pig	Preference for head and face	None
Trimenopon jenningsi	Guinea pig	General body surface	None
ORDER ACARINA (MITES)			
Laelaps sp.	Rat, mouse	Skin during feeding	None
Myobia musculi *Radfordia ensifera*	Rat, mouse	Skin surface	Mange, head and face, particularly in older animals
Radfordia affinis *Myocoptes musculinus*	Rat, mouse	Skin surface	Mange, shoulders and back, particularly in older animals
Psorergates simplex	Mouse	Dermal pockets	Rare. Skin nodules which can become infected. Mange of ears reported once
Chirodiscoides caviae	Guinea pig	Hair shaft	None apparent
Trixacarus caviae	Guinea pig	Epidermal tunnels	Severe mange
Sarcoptes scabiei	All species	Epidermal tunnels	Dry mange with alopecia may be generalized
Notoedres notoedres	Rat	Epidermal tunnels	Mange consists of nodules. Gradual erosion of the edge of the ear
Notoedres muris	Wild rat	Epidermal tissue of ears & face	Erosion of pinna in severe cases
Demodex criceti	Hamster	Dermal pits	None apparent

continued

Appendix A1. *(continued)* Rodents

Species	Host	Location	Pathogenicity
Demodex aurati	Hamster	Hair follicles	Rare, generalized mange in old stock
Demodex merioni	Gerbil	Hair follicles	None apparent
Demodex ratti	Rat	Hair follicles	None apparent
Demodex nanus	Rat	Hair follicles	None apparent
Spelorodens clethrionomys	Hamster	Internal nasal cavity	Only one report. No apparent clinical signs
ENDOPARASITES			
PHYLUM PROTOZOA			
ORDER AMOEBIDA			
Entamoeba muris	Mouse, rat, hamster	Large intestine	None
Entamoeba caviae	Guinea pig	Large intestine	None
ORDER TRICHOMONADIDA			
Trichomonas muris	Rat, mouse	Caecum, colon	None
Trichomonas minuta	Hamster + other rodents	Small intestine	None
Tritrichomonas caviae	Guinea pig	Caecum	None apparent
Trichomonas criceti	Hamster	Caecum	None apparent
Tetratrichomonas sp. and other flagellate sp.	Rat, mouse, hamster, guinea pig	Caecum, colon	None
Chilomastix sp.	Guinea pig	Caecum	None
Spironucleus (Hexamita) muris	Mouse, rat, hamster	Small intestine, caecum	Enteritis in the laboratory mouse
Hexamastix sp.	Rat, hamster	Caecum	None apparent
Retortamonas sp.	Guinea pig, rat, rabbit	Caecum	None apparent
Giardia muris	Rat, mouse, hamster	Small intestine	Enteritis in young mice
Giardia caviae	Guinea pig	Small intestine	None apparent
SUBORDER TRYPANOSOMATINA			
Trypanosoma lewisi	(1) Rat (2) Flea	(1) Blood (2) Alimentary tract	(1) Low; may cause mild anaemia
Trypanosoma duttoni	(1) Mouse (2) Flea	(1) Blood (2) Alimentary tract	(1) Low
CLASS CILIATA			
Balantidium caviae	Guinea pig	Caecum	None
Cyathodinium sp.	Guinea pig	Caecum	None
ORDER EUCOCCIDIORIDA			
Eimeria nieschulzi	Rat	Small intestine	Diarrhoea in young animals dependent on infective dose
Eimeria miyairii	Rat	Small intestine	Mild hyperaemia
Eimeria separata	Rat	Caecum, colon	Mild hyperaemia
Eimeria falciformis	Mouse	Caecum	Mild; animals can survive massive infection
Eimeria pragensis	Mouse	Caecum, colon	Severe. Moderately heavy infection causes diarrhoea, sloughing of mucosa and death
Eimeria vermiformis	Mouse	Caecum, colon	Moderate. Heavy infection causes diarrhoea
Eimeria sp. (10 recorded)	Mouse	Mostly intestinal	Largely unrecorded

continued

Appendix A1. *(continued)* Rodents

Species	Host	Location	Pathogenicity
Eimeria caviae	Guinea pig	Caecum, colon	Heavy infections produce severe effects on young animals; oedema of mucosa, diarrhoea, emaciation and death
Isospora ratti	Rat	Small intestine	None apparent
Toxoplasma gondii	(1) Cat (2) Range of mammalian sp.	(1) Small intestine (2) Brain, eye, cardiac and skeletal muscle	(1) Mild (2) Not well documented
Sarcocystis sp.	Range of rodent species	Muscle fibres	None apparent
Cryptosporidium muris	Mouse	Surface of epithelial cells of stomach	Mild digestive disturbance
Cryptosporidium parvum	Mouse	Epithelial cells of small intestine	None
Klossiella cobayae	Guinea pig	Endothelial cells of kidney tubules	Mild nephritis
Klossiella muris	Mouse	Endothelial cells of kidney tubules	Mild nephritis
Klossiella caviae	Guinea pig	Kidney tubules	None apparent
Plasmodium berghei *Plasmodium vinckei* *Plasmodium chabaudi*	(1) Tree rat (2) Mosquito	(1) Red blood cells	(1) None apparent in natural hosts; fatal malaria in experimental model
Babesia microti	(1) Rodents (2) Tick	(1) Red blood cells	(1) None apparent in natural hosts; malarial experimental model
Babesia decumani	(1) Rat (2) Tick	(1) Red blood cells	
Babesia rodhaini	(1) Tree rat (2) Tick	(1) Red blood cells	
Encephalitozoon (Nosema) cuniculi	Range of rodents	Brain and kidney	None apparent; pseudocysts in brain, perivascular cuffing
ORDER RICKETTSIALES			
Eperythrozoon coccoides	(1) Mouse (2) Blood-sucking arthropod, usually louse	(1) Surface of red blood cells (2) Cycle unknown	(1) Mild anaemia in splenectomized animals
Haemobartonella muris	(1) Rat (2) Blood-sucking arthropod, usually louse	(1) Folds on the surface of red blood cells (2) Cycle unknown	(1) Severe anaemia in splenectomized animals
PHYLUM PLATYHELMINTHES CLASS CESTODA			
Hymenolepis nana	(1) Mouse, rat, hamster (2) Flour beetle or definitive host	(1) Small intestine (2) Body cavity or villus	(1) Varies with density of infection from no apparent clinical disease to chronic inflammation, weight loss in hamster, impaction and death

continued

Appendix A1. *(continued)* Rodents

Species	Host	Location	Pathogenicity
Hymenolepis diminuta	(1) Rat (2) Flour beetle	(1) Small intestine (2) Body cavity	(1) None
Hymenolepis microstoma	(1) Mouse (2) Flea	(1) Duodenum and bile ducts (2) Body cavity	(1) Inflammation, distension of bile ducts
Cataenotaenia pusilla	(1) Mouse (2) Mite *Glycophagus domesticus*	(1) Small intestine (2) Body cavity	(1) None
Cataenotaenia oranesis	(1) Gerbil (2) Mite	(1) Small intestine (2) Mesentery	None apparent in either host
Taenia taeniaeformis/ Cysticercus fasciolaris	(1) Cat (2) Mouse	(1) Small intestine (2) Mesentery	Low in both hosts depending on level of infection
CLASS TREMATODA			
Schistosoma japonicum	Rat	Portal veins	None apparent
PHYLUM ACANTHOCEPHALA			
Moniliformis moniliformis	(1) Rat (2) Cockroach	(1) Intestine (2) Haemocoele	(1) Depends on numbers; can produce ulceration
PHYLUM NEMATODA			
ORDER ASCARIDORIDA			
Syphacia sp.	Rat, mouse, hamster	Caecum. Ova around anus	None apparent
Aspiculuris tetraptera	Mouse	Caecum, colon	None apparent
Heterakis sp.	Rat	Caecum, colon	None apparent
Paraspidodera uncinata	Guinea pig	Caecum	None apparent
Dentostomella translucida	Gerbil	Intestine	None apparent
ORDER STRONGYLORIDA			
Heligmosomoides (Nematospiroides) dubiuus	Mouse	Lower small intestine	Inflammatory reaction to heavy infection
Nippostrongylus brasiliensis	Rat	Small intestine	Inflammation of skin and lungs (larva), and intestine (adult)
Trichostrongylus sp.	Range of rodents	Small intestine	Depends on numbers; enteritis, pneumonia
ORDER DORYLAIMORIDA			
Trichuris muris	Rat, mouse	Large intestine	Some inflammation, depending on level of infection
Capillaria hepatica	Range of wild rodents	Liver	Liver necrosis, cirrhosis
Trichosomoides crassicauda	Rat	Wall of urinary bladder	Low; eventually may produce granulomatous lesions

Appendix A2. Some ecto- and endoparasites of lagomorphs. (1) = Primary host; (2) = Secondary host

Species	Host	Location	Pathogenicity
ECTOPARASITES			
ORDER SIPHONAPTERA (FLEAS)			
Spilopsyllus cuniculi	Rabbit	Body surface; usually found feeding in clumps inside pinna. Larvae in nest	None apparent. Vector of myxoma virus
ORDER PHTHIRAPTERA (LICE)			
Haemodipsus ventricosus	Rabbit	Body surface	None
ORDER ACARINA (MITES)			
Listrophorus gibbus	Rabbit	Hair shaft	None
Cheyletiella parasitivorax	Rabbit	Skin surface	Occasionally mange of shoulders and back
Psoroptes cuniculi	Rabbit	Skin surface of ear canal	Ear canker. Infection can extend to face
Sarcoptes scabiei	All species	Epidermal tunnels	Dry generalized mange, alopecia
Notoedres cuniculi	Rabbit	Epidermal tunnels	Dry generalized mange, alopecia
ENDOPARASITES			
PHYLUM PROTOZOA			
ORDER TRICHOMONADIDA			
Eutrichomastix sp.	Rabbit	Caecum	None
Giardia duodenalis	Rabbit	Small intestine	None apparent
Enteromonas sp.	Rabbit	Caecum	None apparent
Retortamonas sp.	Rabbit	Caecum	None apparent
ORDER EUCOCCIDIORIDA			
Eimeria perforans *Eimeria irresidua* *Eimeria magna* *Eimeria media*	Rabbit	Small intestine	Varying in pathogenicity. *E. irresidua* & *E. magna* most pathogenic, *E. perforans* & *E. media* less so; diarrhoea, enteritis
Eimeria elongata *Eimeria neoloporis*	Rabbit	Intestine	Not known
Eimeria intestinalis	Rabbit	Ileum, caecum	Severe
Eimeria caecicola	Rabbit	Ileum, caecum	Mild
Eimeria piriformis	Rabbit	Ileum, caecum	Unknown
Eimeria stiedae	Rabbit	Bile ducts	Severe, depending on level of infection
Toxoplasma gondii	(1) Cat (2) Range of mammalian sp.	(1) Small intestine (2) Brain, eye, cardiac and skeletal muscle	(1) Mild (2) Not well documented
Sarcocystis sp.	(1) Cat, dog (2) Range of rodents and lagomorphs	(1) Small intestine (2) Cardiac and skeletal muscle	(1) Usually none (2) None apparent
Encephalitozoon (Nosema) cuniculi	Rabbit, rodents	Majority of major organs during acute phase; lesions localized in brain and kidney	Usually inapparent. Neurological symptoms reported in young rabbits heavily infected

continued

Appendix A2. *(continued)* Lagomorphs

Species	Host	Location	Pathogenicity
PHYLUM PLATYHELMINTHES			
CLASS CESTODA			
Cittotaenia ctenoides	(1) Rabbit	(1) Small intestine	(1) Heavy infestation causes digestive disturbance
	(2) Oribatid mites	(2) Body cavity	(2) None
Cysticercus pisiformis	(1) Dog and cat	(1) Small intestine	(1) Usually none
	(2) Rabbit	(2) Peritoneal cavity	(2) Usually none
Multiceps serialis	(1) Dog	(1) Small intestine	(1) None
	(2) Rabbit, hare and wild rodents rare	(2) Subcutaneous tissues	(2) Swelling, puffed skin
Echinococcus sp.	(1) Dog	(1) Small intestine	(1) None
	(2) Rabbits, rodents herbivores and man	(2) Liver, lungs and peritoneal cavity	(2) Often extensive pulmonary or neural effects
PHYLUM NEMATODA			
Passalurus ambiguus	Rabbit	Caecum	Low, depending on numbers
Trichostrongylus retortaeformis	Rabbit	Small intestine	Depending on numbers, penetrate mucosa causing desquamation and inflammation
Graphidium strigosum	Rabbit	Stomach, intestine	
Obeliscoides cuniculi	Rabbit	Small intestine	

Appendix A3. Some ecto- and endoparasites of some New World monkeys. (1) = Primary host; (2) = Secondary host

Species	Host	Location	Pathogenicity
ECTOPARASITES			
ORDER PHTHIRAPTERA (LICE)			
Suborder Anopleura			
Pedicinus mjoebergi	*Ateles* sp., *Cacajao* sp., *Pithecia* sp.	Pelage	None
Pediculus lobatus	*Alouatta* sp., *Ateles* sp.	Pelage	None
Pediculus humanus	*Ateles paniscus*	Pelage	None
Harrisonia uncinata	*Callithrix* sp.	Pelage	None
Suborder Amblycera			
Gliricola pintoi	*Leontideus nigricollis*	Pelage	None
Trichodectus armatus	*Brachyteles* sp.	Pelage	None
Trichodectes semiarmatus	*Alouatta* sp.	Pelage	None
Tetragyropus aotophilus	*Aotus boliviensis*	Pelage	None
ORDER ACARINA (MITES & TICKS)			
Pneumonyssoides stammeri	*Alouatta caraya*	Large bronchioles	None
	Lagothrix lagotricha	Larynx	None
Ixodes loricatus	*Ateles* sp.	Pelage	None
Kutzercoptes grunbergi	*Cebus* sp.	Skin	Irritation
Rhinodex baeri *Stomatodex galagoensis* *Demodex saimiri*	Range of species	Skin	Not known
ORDER PENTASTOMIDA			
Linguatula serrata	(1) Dogs (2) *Saguinus* sp., *Saimiri* sp., *Callicebus* sp., other mammals	(1) Nasal passages (2) Liver, mesentery & lungs	(1) None (2) No apparent clinical signs
Porocephalus clavatus	(1) Snake (2) *Callithrix* sp., *Saguinus* sp.	(1) Airways (2) Liver, mesentery & lungs	(2) No apparent clinical signs
Armillifer armillatus	(1) Snake (2) *Cebus* sp.	(1) Lungs (2) Peritoneal cavity and beneath liver capsule	(2) No apparent clinical signs
ENDOPARASITES			
PHYLUM PROTOZOA			
ORDER AMOEBIDA			
Iodamoeba buetschlii	*Cebus* sp.	Large intestine	None
Entamoeba coli	*Callithrix jacchus*	Large intestine	None
Entamoeba histolytica	Range of species	Colon	Dysenteric amoebiasis in some species
Endolimax nana	Range of species	Caecum, colon	None apparent
Endolimax kueneni	*Cebus* sp.	Large intestine	None
ORDER TRICHOMONADIDA			
Trichomonas sp.	Range of species	Small intestine	None
Giardia lamblia	Range of species	Small intestine	None, except when under stress
Hexamita pitheci	Rhesus monkey	Caecum & colon	None apparent
Embadomonas sp.	Range of species	Small intestine	None

continued

Appendix A3. *(continued)* New World Monkeys

Species	Host	Location	Pathogenicity
Chilomastix sp.	Range of species	Small intestine	None
SUBORDER TRYPANOSOMATINA			
Trypanosoma minasense	(1) *Callithrix jacchus* (2) *Glossina* sp.	(1) Bloodstream	(1) None
Trypanosoma rangeli	(1) *Cebus* sp. (2) *Glossina* sp.	(1) Bloodstream	(1) None
Trypanosoma cruzi	(1) *Callithrix penicillata* (2) *Rhodnius prolixus*	(1) Bloodstream (2) Intestine	(1) None, except when under stress (2) None
Leishmania sp.	(1) *Saguinus* sp. (2) *Phlebotomus* sp.	(1) Skin (2) Intestine	(1) None apparent. Reservoir of human infection (2) None
CLASS CILIATA			
Balantidium sp.	*Saimiri* sp.	Large intestine	Balantidiosis (rare)
Balantidium mentichi	*Ateles* sp.	Large intestine	None
Taliaferria clarki	Spider monkey	Caecum	None apparent
ORDER EUCOCCIDIORIDA			
Isospora arctopitheci	*Saguinus* sp., capuchin	Alimentary tract	None
Isopora callimico	*Callimico goeldii*	Alimentary tract	None
Isopora scorsai	*Cacajao tubicundus*	Alimentary tract	None
Toxoplasma gondii	*Saimiri* sp., *Cebus* sp., *Alouatta* sp.	Brain cysts	None documented
	Oedipomidas oedipus	Acute generalized parasitaemia	One reported fatality
Sarcocystis sp.	*Ateles* sp., *Cebus* sp.	Muscle	None
Plasmodium brasilianum	(1) *Saimiri* sp. (2) Mosquito vector	(1) Red blood cells (2) Salivary glands	(1) Quartan malaria (2) None
Plasmodium simium	(1) *Ateles* sp., *Alouatta* sp., *Cebus* sp. (2) Mosquito vector	(1) Red blood cells (2) Salivary glands	(1) Tertian malaria (2) None
Babesia sp.	(1) *Callithrix jacchus* (2) Unknown vector	(1) Red blood cells	(1) Babesiosis
Encephalitozoon (Nosema) cuniculi	*Saimiri* sp.	Brain cysts	Seizures
Nosema sp.	*Callicebus moloch*	Small intestine	None documented
PHYLUM PLATYHELMINTHES CLASS CESTODA			
Atriotaenia sp.	Squirrel monkey	Intestine	None apparent
Sparganum	Rhesus monkey	Connective tissue	None apparent
CLASS TREMATODA			
Schistosoma mansoni	*Saimiri sciureus*	Mesenteric veins	Diarrhoea with blood and mucus. Mortality depends on level of infection (rare)
PHYLUM ACANTHOCEPHALA			
Prosthenorchis elegans *Prosthenorchis spirula*	Range of species	Ileum to rectum	Inflammation & ulceration; peritonitis; can cause death
PHYLUM NEMATODA ORDER ASCARIDORIDA			
Lobatorobius scleratus	*Saimiri* sp.	Large intestine	None

continued

Appendix A3. *(continued)* New World Monkeys

Species	Host	Location	Pathogenicity
Trypanoxyuris trypanuris	*Saguinus* sp., *Callithrix* sp.	Large intestine	None
Trypanoxyuris ateles	*Ateles* sp.	Large intestine	None
Enterobius vermicularis	*Cebuella* sp.	Large intestine	None
Enterobius microon	*Cebuella* sp.	Large intestine	None
Enterobius interlabiatus *(Trypanoxyuris lobata)*	*Callithrix* sp., *Oedipomidas* sp., *Aotus* sp.	Large intestine	None
Subulura jacchi	*Callithrix* sp.	Large intestine	None
ORDER SPIRURORIDA			
Protospirura muricola	*Cebus* sp.	Stomach	Mechanical blockage
Physaloptera dilatata	Capuchin, *Callithrix* sp.	Stomach	Ulceration
Gongylonema saimiri	*Callicebus* sp., *Saimiri* sp.	Stomach and oesophageal wall	None
Trichospirura leptostoma	*Saguinus oedipus*, *Callithrix jacchus*, *Aotus* sp., *Callimico goeldii*	Pancreatic duct	Pancreatic dysfunction
Dipetalonema gracili	(1) *Saimiri sciureus*	(1) Adults in serous cavities. Microfilariae in blood	(1) May produce inflammatory response and adhesions
	(2) Arthropod	(2) Not documented	
Dipetalonema caudispina	(1) *Saimiri* sp.	(1) Adults in abdominal cavities and intermuscular fascia. Microfilariae in blood	(1) None
	(2) Arthropod	(2) Not documented	
Dipetalonema parvum	(1) *Saimiri* sp., Capuchins	(1) Adults in the connective tissue. Microfilariae in blood	(1) None
	(2) Arthropod	(2) Not documented	
Dipetalonema obtusa	(1) *Cebus* sp., *Saimiri* sp.	(1) Perioesophageal connective tissue. Microfilariae in blood	(1) Unknown
	(2) Arthropod	(2) Not documented	
Tetrapetalonema panamensis *Tetrapetalonema tamarensis* *Tetrapetalonema mystaxi*	*Cebus* sp., *Saguinus* sp.	Pleural cavities and subcutaneous connective tissue	None
Tetrapetalonema saimiri	*Saimiri sciureus*	Subcutaneous connective tissue	None
Tetrapetalonema atelensis	*Ateles geoffroyi*	Subcutaneous connective tissue	None
Filaria intercostatis	*Saimiri* sp.	Intercostal	None
ORDER STRONGYLORIDA			
Oesophagostomum sp.	*Saimiri* sp.	Large intestine	Rare. Nodular cysts in submucosa. Secondary bacterial infection may cause peritonitis

continued

Appendix A3. *(continued)* New World Monkeys

Species	Host	Location	Pathogenicity
Molineus elegans	*Saimiri* sp., *Aotus* sp., *Cebus* sp.	Small intestine	None
Molineus vexillarius	*Saguinus* sp.	Duodenum pylorus	Nodular lesions of serosal surface
Molineus torulosus	*Saimiri* sp., *Aotus* sp., *Cebus* sp.	Small intestine	Haemorrhage, ulceration
Longistriata dubia	*Saimiri* sp.	Small intestine	Unknown
Rictularia sp.	*Saguinus* sp., capuchin, *Callithrix* sp.	Small intestine	Unknown
Filaroides barretoi	*Saimiri sciureus*, *Callithrix jacchus*	Lung	Pigmented lung nodules. No apparent disease
Filaroides gordius	*Saimiri* sp.	Lung	Pigmented lung nodules. No apparent disease
Filaroides cebus	*Cebus* sp.	Lung	Pigmented lung nodules. No apparent disease
Filaroides cebuellae	*Cebuella pygmaea*	Lung	Pigmented lung nodules. No apparent disease
ORDER RHABDITORIDA			
Strongyloides cebus	*Cebuella pygmaea*; range of other species	Small intestine, muscularis-mucosae	Enteritis
ORDER DORYLAIMORIDA			
Capillaria hepatica	*Ateles* sp.	Liver	Nodules and scarring of liver capsule

Appendix A4. Some ecto- and endoparasites of some Old World monkeys. (1) = Primary host; (2) = Secondary host

Species	Host	Location	Pathogenicity
ECTOPARASITES			
ORDER SIPHONAPTERA (FLEAS)			
Ctenocephalides canis *C. felis strongylus*	Range of species	General body surface	Irritation
Tunga penetrans	Gorilla, baboon	Skin	Burrows deep into the epidermis and provides a focus for bacterial infection
ORDER PHTHIRAPTERA (LICE)			
Suborder Anopleura			
Pedicinus sp.	Range of species	Pelage	Irritation, pruritis
Pediculus sp.	Man, great apes	Pelage	Unknown
Phthirus sp.	Man, great apes	Pelage	Unknown
ORDER ACARINA (MITES & TICKS)			
Sarcoptes sp.	*Papio* sp.	Epidermal tunnels, skin	Mange
Prosarcoptes pitheci	*Papio papio*, *Cercopithecus* sp.	Skin	Mange
Prosarcoptes faini	Chacma baboon	Skin	Mange
Prosarcoptes scanloni	Macaques	Skin	Mange
Paracoroptes gordonii	*Papio* sp.	Skin	Mange
Kutzercoptes cercopitheci	*Cercopithecus mona mona*	Skin	Mange
Psorergates sp.	Range of species	Skin	Mange
Pneumonyssus sp.	Macaques, *Cercopithecus* sp., *Papio* sp.	Lung, in nodules	Cough, otherwise few clinical signs
Rhinophaga sp.	Rhesus, guenons	Lungs	Unknown
Rhinophaga papionis	Baboon	Lungs	Pneumonitis
Amblyomma sp.	*Papio* sp.	Young and adult stages found attached to skin for various periods during development	Irritation during feeding. Vectors of bloodborne diseases
Riphicephalus sp.	*Papio* sp.	Pelage	Irritation during feeding
ORDER PENTASTOMIDA			
Armillifer armillatus *Pentastomium* sp. *Porocephalus* sp.	(1) Python and other snakes (2) *Papio* sp., macaques, mandrills	(1) Respiratory tract (2) Liver & mesentery	(2) None (2) No clinical signs
ENDOPARASITES			
PHYLUM PROTOZOA			
ORDER AMOEBIDA			
Dientamoeba fragilis *Endolimax nana* *Entamoeba coli* *Entamoeba chattoni*	Wide range of species	Large intestine	None
Entamoeba gingivalis	Range of macaques	Mouth	None
Entamoeba hartmanni	Range of species	Large intestine lumen	None
Entamoeba histolytica	Chimpanzee, orang-utan, macaques	Large intestine lumen, ulcer and liver abscess	Vomiting, diarrhoea with blood and mucus, wasting and loss of appetite

continued

Appendix A4. *(continued)* Old World Monkeys

Species	Host	Location	Pathogenicity
Endolimax nana	Range of species	Colon	None
Iodamoeba buetschlii	Range of species	Colon	None
ORDER TRICHOMONADIDA			
Chilomastix mesnili	Range of species	Large intestine	None
Giardia lamblia	Range of species	Duodenum	None
Hexamita pitheci	*Macaca mulatta*	Small intestine	None
Trichomonas tenax	*Macaca mulatta*, *M. fascicularis*	Mouth	None
Trichomonas sp.	*Macaca mulatta*, *M. fascicularis*	Small intestine	None
Trichomitus wenyoni	Rhesus monkey, baboon	Caecum, colon	None
Pentatrichomonas hominis } *Retortamonas intestinalis*	*Macaca mulatta*, *M. fascicularis*	Small intestine	None
CLASS CILIATA			
Balantidium coli	Old World species	Colon	None
Troglodytella sp.	Chimpanzee, gorilla	Colon	One report of diarrhoea and death
ORDER EUCOCCIDIORIDA			
Cryptosporidium sp.	*Macaca mulatta*	Small intestine	Mild diarrhoea, villous atrophy
Isospora papionis	*Papio ursinus*	Small intestine	None apparent
Toxoplasma sp.	(1) Cat (2) Range of mammalian sp.	(1) Small intestine (2) Brain, eye, cardiac and skeletal muscle	(1) Mild (2) Not well documented
Sarcocystis kortei	(1) *Macaca mulatta*, *Erythrocebus* sp., *Cercopithecus* sp., (2) Not known	(1) Skeletal muscle	(1) None
Sarcocystis nesbitti	(1) *Macaca mulatta* (2) Not known	(1) Skeletal muscle	(1) None
Plasmodium cynomolgi	(1) Chiefly *Macaca* sp. (2) Mosquito	(1) Blood cells (2) Salivary glands	(1) Tertian malaria, mild fever, splenomegaly, hepatic congestion
Plasmodium knowlesi	As above	As above	Quotidian malaria, severe in experimental infection
Plasmodium inui	As above	As above	Quartan malaria, mild
Plasmodium coatneyi	As above	As above	Tertian malaria, mildly pathogenic
Plasmodium fieldi	As above	As above	Tertian malaria, mild
Plasmodium semiovale	As above	As above	Tertian malaria, mild
Plasmodium fragile	*Macaca sinica*, other macaques	As above	Tertian malaria. Reportedly high mortality
Hepatocystis kochi	(1) Africa green monkeys; type host *Cercopithecus* sp. (2) *Culicoides* (midge)	(1) Blood cells, liver tissue	(1) No apparent disease
Hepatocystis semnopitheci	(1) *Presbytis entellus* (2) *Culicoides* sp.	(1) Blood cells, liver tissue (2) Thoracic muscle	(1) None
Babesia pitheci	(1) Macaques, *Pan* sp., *Papio* sp. (2) Tick	(1) Red blood cells	(1) Anaemia, fever

continued

Appendix A4. *(continued)* Old World Monkeys

Species	Host	Location	Pathogenicity
Entopolypoides macaci	(1) *Cercopithecus* sp., macaques (2) Not known	(1) Red blood cells	(1) Mild anaemia
Pneumocystis carinii	Widespread in many species including primates, e.g. chimpanzee	Lung	Fatal in immunodepressed animal
PHYLUM PLATYHELMINTHES CLASS CESTODA			
Multiceps sp.	(1) Adults in dogs and other carnivores	(1) Adults in lumen of small intestine	(1) None
	(2) *Papio* sp., macaques	(2) Coenurus in subcutaneous tissues	(2) Rarely, only when central nervous system involved
Echinococcus granulosus	(1) Adult in dog, fox and relatives	(1) Adults in lumen of large intestine	(1) None
	(2) All mammalian species vulnerable	(2) Hydatid in lungs, liver and other parts of the body	(2) Varying with situation. Tumour-like symptoms
Taenia hydatigena	(1) Adults in dogs	(1) Adult in small intestine	(1) None
	(2) *Papio* sp., macaques	(2) *Cysticercus tenuicollis* in peritoneal cavity	(2) None
Taenia solium	(1) Pig	(1) Adult in small intestine	(1) None
	(2) *M. cyclopis*, *Cercopithecus* sp.	(2) Musculature and subcutaneous tissues	(2) None
Bertiella sp.	(1) Wide variety, e.g. macaques, *Papio* sp., *Hylobates* sp., *Pan* sp., *Pongo* sp.	(1) Adults in small intestine	(1) None
	(2) Oribatid mites	(2) Cysticercoid in body cavity	(2) None
Hymenolepis sp.	(1) *Hylobates* sp.	(1) Adult in small intestine	(1) None
	(2) Grain beetle	(2) Cysticercoid in body cavity	(2) None
Spirometra sp.	(1) Adults in variety of vertebrates and invertebrates	(1) Adults in intestinal lumen	(1) None
	(2) *Papio* sp., *Macaca fascicularis*, *Theropithecus* sp.	(2) Sparganum in connective tissue	(2) Inflammation during migration
Mesocestoides sp.	(1) Adults in birds and other mammals	(1) Adults in intestinal lumen	(1) None
	(2) *Papio* sp., macaques, gorilla	(2) Tetrathydria in body cavity	(2) None

continued

Appendix A4. *(continued)* Old World Monkeys

Species	Host	Location	Pathogenicity
CLASS TREMATODA (except where specified, all trematodes have an intermediate mollusc host)			
Schistosoma mansoni	*Papio anubis*, many other sp.	Portal vein, mesenteric vein	Diarrhoea with blood and mucus at the time of egg laying. Mortality depends on level of infection
Schistosoma haematobium	Many species vulnerable; *Cercopithecus sabaeus*, *Papio* sp., macaques	Pelvic and visceral veins	Haematuria; thickened bladder wall
Schistosoma japonicum	Macaques	Portal vein, branches of mesenteric vein	Haemorrhagic diarrhoea; thickened intestinal wall
Schistosoma mattheei	*Papio* sp.	Portal vein	Cirrhosis
Fasciola hepatica	Macaques	Bile duct	Liver enlarged and fibrosed
Brodenia sp.	*Cercocebus* sp., *Papio* sp.	Pancreas	Unknown
Dicrocoelium dendriticum	Macaques, *Cercopithecus* sp., *Papio* sp.	Bile ducts	Liver enlarged with bile duct obstruction
Eurytrema satoi	Macaques	Bile ducts	Liver enlarged with bile duct obstruction
Opisthorchis sp.	Cynomolgus	Bile and pancreatic ducts	Depending on level of infection cause catarrhal inflammation
Paragonimus sp.	(1) Macaques (2) Mollusc (3) Crustacean	(1) Lungs or liver (2) Viscera (3) Metacercariae in muscle	(1) Respiratory distress
Gastrodiscus hominis	(1) Macaques (2) Mollusc	(1) Caecum, colon (2) Viscera	Mucoid, mild diarrhoea, chronic enteritis
Watsonius sp.	(1) Guenons, *Papio* sp. (2) Mollusc	(1) Intestine (2) Viscera	Mucoid, mild diarrhoea, chronic enteritis
Clonorchis noci	(1) Macaques (2) Mollusc	(1) Intestine (2) Viscera	Low
Ogmocotyle indice	(1) Macaques (2) Mollusc	(1) Intestine (2) Viscera	Low
Phaneropsolus oviformis	(1) *Nycticebus* sp. (2) Mollusc	(1) Intestine (2) Viscera	Low
Primatotrema macacae	(1) Macaques (2) Mollusc	(1) Intestine (2) Viscera	Low
PHYLUM NEMATODA			
ORDER ASCARIDORIDA			
Enterobius sp.	Macaques, baboons, apes	Large intestine	Occasional perianal pruritis
Oxyuris armata	*Papio hamadryas*	Colon	None
Subulura sp.	Macaques, *Cercopithecus* sp., mandrill	Colon	None
Ascaris lumbricoides	Macaques, gibbon, apes	Small intestine	Usually not pathogenic, but reported fatality in chimpanzee
ORDER SPIRURORIDA			
Streptopharagus sp.	Macaques, *Erythrocebus patas*, gibbons, baboons	Stomach	None

continued

Appendix A4. *(continued)* Old World Monkeys

Species	Host	Location	Pathogenicity
Gongylonema sp.	Rhesus monkeys, baboons	Lungs, tongue, oesophagus	None
Physocephalus sp.	Rhesus monkeys, baboons	Stomach	None
Physaloptera tumefaciens	Macaques	Stomach	Gastritis, haemorrhage
Physaloptera caucasica	*Papio* sp.	Stomach	Gastritis, haemorrhage
Dipetalonema gracili	(1) Macaques	(1) Adult—oesophagus, body cavities; larvae—bloodstream	(1) None known
	(2) Arthropod	(2) Not documented	
Dipetalonema aethiops	(1) *Cercopithecus aethiops*	(1) Adult—inter-fasicular connective tissue; larvae—bloodstream	(1) None known
	(2) Arthropod	(2) Not documented	
Dirofilaria sp.	(1) Monkeys, apes	(1) Adult—heart, tissues; larvae—bloodstream	(1) Mechanical blockage; depends on number of worms
	(2) Mosquito	(2) Larvae in thoracic muscles	
Edesonfilaria malayensis	(1) *Macaca fascicularis*	(1) Adult—peritoneum; larvae peripheral blood-stream	(1) None
	(2) Arthropod	(2) Unknown	
Onchocerca volvulus	(1) Baboons	(2) Adult—sub-cutaneous; larvae—cutaneous	(1) None
	(2) Simulium	(2) Larvae in thoracic muscles	
Loa loa	(1) *Papio* sp., mangabey, *Cercopithecus* sp.	(1) Adult—sub-cutaneous body cavities & muscle; larvae—blood-stream	(1) Nodular lesions in spleen
	(2) Chrysops	(2) Larvae in thoracic muscles	
Brugia malayi } *Brugia pahangi* }	(1) Macaques	(1) Adults in lymphatics; micro-filariae in blood	(1) None apparent
	(2) Mosquito	(2) Larvae in thoracic muscles	
ORDER STRONGYLORIDA			
Oesophagostomum sp.	Macaques, baboons, chimpanzee, gorilla	Adults in colon	Nodules. Haemorrhage in colon in heavy infections
Ternidens diminutus	Variety of guenons and macaques	Caecum, colon	Nodules in colon, anaemia
Ancylostoma duodenale	Macaques	Small intestine	Severe anaemia, abdominal symptoms
Necator americanus	*Erythrocebus patas*, mandrill, *Papio* sp., apes	Small intestine	Unknown. May cause anaemia as in man

continued

Appendix A4. *(continued)* Old World Monkeys

Species	Host	Location	Pathogenicity
Trichostrongylus sp.	Macaques, *Papio* sp.	Small intestine	None
Nochtia nochti	Macaques	Stomach	Can induce gastric tumours
ORDER RHABDITORIDA			
Strongyloides sp.	*Papio* sp., macaques, apes	Small intestine	Diarrhoea, anaemia
ORDER DORYLAIMORIDA			
Anatrichosoma sp.	Macaques, *Papio* sp., *Erythrocebus patas*	Nasal epithelium	None
Trichinella spiralis	Many species including primates	Adults—mucosa of small intestine; larvae—musculature	Adult—mild enteritis. Larvae—flu symptoms during migratory phase
Trichuris trichiura	Macaques, baboons, apes	Caecum, colon	Usually none; rarely diarrhoea
Capillaria hepatica	Monkeys, apes	Liver	Rarely hepatitis, cirrhosis

Appendix A5. Zoonotic organisms of primates

Organism	Primate host other than man
ECTOPARASITES	
PHYLUM ARTHROPODA	
Order Siphonaptera	
Tunga penetrans	Gorilla, baboon
Order Diptera	
Cuterebra sp.	Howler monkey, *Alouatta* sp.
Cordylobia sp.	
Suborder Anopleura	
Pediculus humanis capitis	Red-faced black spider monkey, *Ateles paniscus*
Order Acarina	
Family Trombiidae	Wide range of Asian monkeys
Family Sarcoptidae (*Sarcoptes scabiei*)	Wide range of both Old and New World monkeys
Order Pentastomida	
Family Linguatulidae (*Armillifer armillatus*)	Wide range of both Old and New World monkeys
ENDOPARASITES	
PHYLUM PROTOZOA	
Class Rhizopoda	
Entamoeba histolytica	Majority of Old and New World monkeys
Entamoeba chattoni	*Pan troglodytes*, *Macaca mulatta*
Class Mastigophora	
Giardia lamblia	*Papio cynocephalus*, *Pan troglodytes*
Trichomonas sp. *Chilomastix* sp. *Embadomonas* sp.	Range of Old and New World species with apparently similar species of flagellate
Class Ciliata	
Balantidium coli	*Pan troglodytes*, macaques
Class Sporozoasida	
Plasmodium schwetzi	Chimpanzee
Plasmodium (rhodhaini) malariae	Chimpanzee
Plasmodium knowlesi	Macaque
Plasmodium simium	*Alouatta* sp.
Plasmodium brasilianum	*Saimiri* sp.
Plasmodium cynomolgi	*Macaca fascicularis*
Toxoplasma gondii	Wide range of species
PHYLUM PLATYHELMINTHES	
Class Cestoda	
Bertiella sp.	Wide range of Old and New World species
Hymenolepis nana	*Callicebus* sp., *Cercopithecus nictitas*, *Pan troglodytes*, *Saimiri sciureus*
Taenia solium	*Papio* sp.
Multiceps serialis	*Theropithecus gelada*
Class Trematoda	
Gastrodiscus hominis	*Macaca mulatta*, *Macaca fascicularis*, *Papio* sp., *Pan* sp., *Cercocebus* sp., *Cercopithecus* sp., *Saimiri sciureus*
Watsonius watsoni	*Papio* sp., *Cynopithecus* sp., *Cercopithecus* sp., *Callithrix* sp.
Fasciolopsis buski	*Macaca mulatta*, *Macaca fascicularis*
Fasciola hepatica	*Macaca mulatta*
Clonorchis sinensis	*Macaca cyclopis*, *Macaca fascicularis*
Paragonimus westermanni	*Macaca fascicularis*
Dicrocoelium dendriticum	Macaques
PHYLUM NEMATODA	
Capillaria hepatica	*Macaca mulatta*, *Cebus* sp., *Ateles* sp., *Pan* sp.

continued

Appendix A5. *(continued)* Zoonotic organisms

Organism	Primate host other than man
Strongyloides stercoralis	*Pan troglodytes*, *Hylobates* sp.
Strongyloides fulleborni	*Pan troglodytes*
Ternidens diminutus	*Cercopithecus* sp., macaques
Ancylostoma duodenale	Macaques, *Hylobates* sp., gorilla
Necator americanus	*Erythrocebus* sp., *Mandrillus sphinx*, *Papio* sp., *Lagothrix* sp., *Ateles* sp., *Pan* sp., gorilla
Oesophagostomum apiostomum	Old World monkeys and apes
Oesophagostomum stephanostomum	Old World monkeys and apes
Ascaris lumbricoides	*Macaca mulatta*, *Hylobates* sp., *Pan* sp., *Pongo* sp., gorilla
Brugia malayi	Macaques, *Nycticebus cougang*, *Presbytis* sp., *Pan* sp., gorilla
Loa loa	Old World baboons (c.f. *Loa papionis*)
Onchocerca volvulus	Chimpanzee
Dracunculus medinensis	Guenon, *Papio* sp.
Anatrichosoma sp.	*Macaca mulatta*, *Macaca fascicularis*, *Cercopithecus* sp., *Erythrocebus* sp., *Cercocebus* sp., *Papio* sp.
Enterobius sp.	Apes, gibbons, marmosets, *Presbytis* sp.

APPENDIX B

HISTOLOGICAL FIXATIVES

Carnoy's fluid

30 ml chloroform
60 ml ethanol
10 ml glacial acetic acid
Fix for approximately one hour then wash three times in 90% ethanol before processing, allowing one hour for each wash.
NB. Ensure specimen samples are no more than 2 mm thick. A thinner piece may be cut after 15 minutes as very thin pieces are likely to curl if fixed directly.

Zenker formalin or Helly's fluid

50 g mercuric chloride
20 g potassium dichromate
5–10 g sodium sulphate
Dissolve in one litre of distilled water. Add one part of 4% formaldehyde to 9 parts of mixture before use.

STAINS

Heidenhain's iron haematoxylin

Mordant
Iron alum, ammonium ferric sulphate, at 3% in distilled water
Stain
10 parts 5% haematoxylin in 96% alcohol
90 parts distilled water
Counterstain
1% eosin in 90% ethanol, or
0.5% orange G in 90% ethanol

Method
(1) Take histological sections to water.
(2) Mordant in 3% iron alum for 30 minutes to 24 hours.
(3) Wash in distilled water.
(4) Stain in haematoxylin for 30 minutes to 24 hours.
(5) Wash rapidly in distilled water.
(6) Differentiate in 1.5% iron alum.
(7) Wash in distilled water.
(8) Counterstain for 1–5 minutes.
(9) Differentiate in 90% ethanol.
(10) Clear and mount.
NB. After fixation with Zenker use the longer time for mordant and stain.

Mayer's haemalum

1.0 g haematoxylin
1000 ml distilled water
50 g ammonium or potassium alum
0.2 g sodium iodate
1.0 g citric acid
50 g chloral hydrate

Method
(1) Dissolve the haematoxylin in the distilled water using heat if necessary.
(2) Add the alum and shake until dissolved; this may also be heated if necessary.
(3) Add the sodium iodate, citric acid and chloral hydrate.
(4) Pour on enough of the stain to cover the smear and gently warm over a spirit light until steam starts to rise, but on no account allow to boil.
(5) Leave aside for 10–15 minutes and rinse in tap water.
(6) Air dry and clear the specimen in xylene and mount in DPX.
This is a good counterstain giving good nuclear staining. It is particularly useful for staining amoebae in sections: in this instance use diluted with an equal quantity of distilled water.

Loeffler's methylene blue

300 ml saturated ethanolic solution of methylene blue
10 ml potassium hydroxide (1% KOH)
990 ml distilled water
This solution may be used as a 1 in 10 dilution with distilled water. Even at this dilution it is a strong and rapid stain. Dip slide with sections into the solution for a few seconds and wash in tap water.

Ehrlich's haematoxylin for helminths

2.0 g haematoxylin
100 ml absolute ethanol
100 ml distilled water
10 ml glacial acetic acid
100 ml glycerol
10 g potassium alum
Allow 6 weeks to ripen. Before use, dilute one part of stock solution with three parts of 45% aqueous acetic acid. The stock solution will keep for up to 6 months.

Method
(1) The time allowed for staining of the fixed specimens varies with size from 2 to 20 minutes, so examine at intervals.
(2) Differentiation with 45% acetic acid for up to 5 minutes can be carried out, but is not very satisfactory.
(3) Dehydrate in glacial acetic acid depending on size; e.g. *Fasciola hepatica* 30 minutes, Acanthocephala 6–10 hours.
(4) Clear with methyl salicylate using, in sequence, 3:1, 1:1 and 1:2 mixtures of acetic acid and salicylate. Specimens will float until penetration is complete.
(5) Mount in canada balsam or Xam.
NB. Haematoxylin is best for trematodes and trichrome for cestodes.

Leishman's stain for thin blood films

(1) Allow a thin blood film to dry in air and place face upwards on a staining rack across a sink.
(2) Fix and stain for 10–30 seconds with 0.5 ml of the undiluted stain.
(3) Carefully add approximately double the quantity of distilled water (1–1.5 ml) buffered to pH 6.7–7.2.
(4) Mix gently and thoroughly by running the suspension in and out of a Pasteur pipette and leave for 8–10 minutes.
(5) Wash by flooding the slide with buffered distilled water.
(6) Differentiate by allowing the buffered water to remain on the slide for 0.5–1 minute.
(7) Rinse off by flooding the slide and leave at an angle to dry.

Field's stain for thick blood films

Obtained as prepared solutions A and B.
(1) Do not fix smear before staining as the blood needs to haemolyse. If the smear is very thick the slide can be briefly dipped into a beaker of distilled water before staining.
(2) Dip into solution A for 5 seconds.
(3) Rinse in tap water for 5 seconds.
(4) Dip into solution B for 3 seconds.
(5) Rinse well in tap water and allow to dry.

Giemsa stain for thick and thin blood films

Stain is usually obtained as a prepared solution but can be prepared as follows:
3.8 g Giemsa powder
250 ml methanol
250 ml glycerine
(1) Grind powder finely in a mortar.
(2) Add glycerine slowly while grinding.
(3) Add half of the methanol to the mortar mix, then pour the mixture into a Pyrex bottle.
(4) Use the remainder of the methanol to rinse out the mortar and pour into the bottle.
(5) Incubate for 24 hours at 37 °C shaking occasionally.
(6) Always filter before use.

Thin films
(1) Fix the film in methanol for 30 seconds.
(2) Add 2 drops of stain from a Pasteur pipette to 1 ml of buffered water (approximately 10% solution).
(3) Stain film face downwards on a staining tray for 20–30 minutes.
(4) Wash slide thoroughly in buffered water.
(5) Allow to dry.

Thick films
NB. Do not fix film before staining.
(1) Place film in staining solution (1 drop from a Pasteur pipette to 1 ml buffered water) and leave for one hour; or if more dilute (35 drops to 100 ml) leave overnight.
(2) Wash thoroughly and gently with buffered water and allow to dry.

Giemsa colophonium method for histological sections containing protozoa (Bray & Garnham 1962)

Buffer
5.0 g sodium dihydrogen phosphate (anhydrous)
2.1 g potassium dihydrogen phosphate
1000 ml distilled water
Giemsa stain
1 part Giemsa solution
1 part acetone
1 part methanol
10 parts buffered water

Method
(1) Bring sections (preferably formalin-fixed, wax-embedded, thickness 2–4 μ) to water.

(2) Stain formalin-fixed sections for 4 hours and Carnoy-fixed sections for 48 hours in Giemsa solution.
(3) Wash quickly in tap water.
(4) Pour on a solution of acetic acid (10 drops from a Pasteur pipette in 250 ml).
(5) Rock slide gently until blue stain starts to come away.
(6) Pour on colophonium resin using a squeezy bottle and rock again until the pink staining is revealed. Add a few drops of tap water while this is happening if the differentiation process needs speeding up.
(7) Dehydrate in acetone/xylene (7:3) solution.
(8) Mount in green euparal.

NB. Disposable gloves are recommended for protection during the differentiation process as the colophonium resin clumps into thick, sticky lumps which can only be removed with the acetone/xylene mixture. The degree of differentiation needs to be assessed by examining the sections during the process. This is not an easy or convenient technique but when mastered will produce excellent results.

For staining of histological sections without using colophonium resin, a clean and crisp result is obtained by differentiating the sections in ethanol with a few drops of 2% acetic acid (approximately 1 drop/10 ml).

Sargeaunt's stain for amoebic cysts (Sargeaunt 1962)

0.2 g malachite green
3.0 ml glacial acetic acid
3.0 ml ethanol
100 ml distilled water

Method

Dissolve the malachite green in the ethanol, add the acetic acid and then the distilled water. This will keep for some time on the bench.

(1) Mix one drop of faecal concentrate (ether sediment) with one drop of stain and examine under a coverslip.
(2) Cyst outlines and nuclear membranes are stained a faint green. Chromatids stain as solid green blocks.

NB. This method will not work directly or after any concentration method. The preliminary action of ether is essential.

Acetic acid alum carmine stain for cestodes

(1) Boil excess carmine stain for 30 minutes in a saturated solution of potash alum, then cool and filter.
(2) To 9 parts of stain add one part of acetic acid.
(3) For use dilute 5–10 times with distilled water.
(4) Stain previously fixed and, where necessary, compressed tapeworms (*see* Techniques) for 1–24 hours depending on the bulk of the segments. Warming gently will accelerate staining.
(5) Decolourize in acid water or acid alcohol.
(6) Dehydrate in ascending grades of ethanol giving plenty of time at each grade.
(7) Clear in cedarwood or clove oil.
(8) Mount in canada balsam.

Horen's trichrome stain for helminths

0.6 g chromotrope 2R C129
0.3 g fast green FCF
0.7 g phosphotungstic acid
1.0 ml glacial acetic acid
100 ml distilled water

For use dilute one part of stock solution to 20 parts of 45% acetic acid. This must be made up freshly before use.

Trichrome stain for intestinal protozoa (Alger 1966)

100 ml distilled water
1.0 ml glacial acetic acid
0.7 g phosphotungstic acid
0.4 g chromotrope 2R
0.3 g light green certified
0.1 g Bismark brown certified

Method

(1) Fix coverslip carrying thin smear of faecal or caecal material in Shaudinn's fluid. Do this by dropping it face downwards on to a shallow layer of the fixative in a petri dish.
(2) After a few minutes remove coverslip into 70% ethanol with iodine to remove mercuric chloride.
(3) Place in clean 70% ethanol for 15 minutes.
(4) Dip coverslip 10 times into 50% ethanol containing 0.1 ml of concentrated hydrochloric acid per 10 ml of ethanol (one second per dip).
(5) Stain for 10 minutes in undiluted stain (can be re-used).
(6) Dip 10 times in 90% ethanol plus 0.1 ml of glacial acetic acid per 10 ml of ethanol.
(7) Dip 10 times in 90% ethanol.
(8) Dip 10 times in 100% ethanol.
(9) Allow to stand in a second dish of 100% ethanol for 3 minutes.
(10) Mount in euparal.

Parasites will be red/green or green. The minute details of the nucleus and cytoplasmic structures (chromidial bars) should be clearly visible. This is a particularly good stain for *Dientamoeba fragilis*.

Gram-Twort's stain for *Encephalitozoon* sp.

Stock solutions

(i) *Aniline crystal violet*:
A: 2 g crystal violet
12 ml absolute ethanol
B: 2 ml aniline oil
98 ml warm distilled water

(ii) *Modified Twort's stain:*
500 ml absolute ethanol
0.1 g fast green
0.9 g neutral red
For use dilute stock solution one part to three of distilled water.

(iii) *Gram's iodine:*
1 g iodine
2 g potassium iodide
300 ml distilled water
Dissolve potassium iodide in 100 ml of distilled water, add the iodine and the remainder of the water.

(iv) *2% alcoholic acetic acid*

Method

(1) Histological sections should be brought to water and well rinsed.
(2) Pour on filtered aniline crystal violet and leave for 3–5 minutes.
(3) Rinse well with distilled water.
(4) Pour on Gram's iodine and leave for 3–5 minutes.
(5) Rinse well with distilled water.
(6) Wash with 2% alcoholic acetic acid until no further stain comes away (check for black crystals).
(7) Pour on diluted Twort's stain and leave for 5 minutes.
(8) Rinse well with distilled water.
(9) Differentiate with 2% alcoholic acetic acid for 15 seconds.
(10) Rinse rapidly in absolute ethanol and xylene.
(11) Leave in xylene for 3 minutes.
(12) Mount in DPX.

Safranin staining method for cryptosporidia (Baxby *et al.* 1984)

(1) Mix a small amount of faecal material with a little water on a glass slide.
(2) Make a smear and allow to dry.
(3) Pass slide over a flame.
(4) Fix in 0.3% hydrochloric acid in methanol for 3–5 minutes.
(5) Add 1% Safranin to the slide and heat until steam is produced and the stain bubbles for one minute (maintain the bubbling for the whole minute).
(6) Wash in distilled water.
(7) Stain in 1% methylene blue for 30 seconds.
(8) Wash in distilled water.
(9) Allow to dry and mount in DPX.

MOUNTANT

Glycerin jelly

10 g gelatin
70 ml glycerin
0.25 g phenol crystals
60 ml distilled water
Dissolve the gelatin in the distilled water using a beaker placed in a hot water bath. Add the glycerin and finally the phenol. This mountant is used for the mounting of small nematodes. When set it can be ringed around with nail varnish and will remain in good condition for several years.

ANTICOAGULANT

Alsever's solution

4.66 g glucose
2.0 g sodium citrate
1.0 g sodium chloride
Dissolve in 200 ml of distilled water and refrigerate.

CONCENTRATION METHODS FOR FAECAL MATERIAL

Flotation method using saturated sodium chloride

(1) Weigh out 1–2 g faecal material.
(2) Homogenize in a little saturated salt (NaCl) solution, using a thin wooden spatula.
(3) Make up liquid to 15 ml with saturated NaCl.
(4) Pour through sieve to remove coarse material.
(5) Place in a centrifuge tube the top of which has been carefully flattened beforehand using emery paper on a flat background.
(6) Top up the tube with saturated NaCl using a squeezy bottle until a convex meniscus is produced.
(7) Carefully lower a coverslip into position on the top of the tube making sure that there are no air bubbles.
(8) Centrifuge at lowest revolutions for 2 minutes.
(9) Remove with a steady upward movement, place on a slide and examine microscopically.

(A comprehensive guide to qualitative and quantitative methods can be found in the *Manual of Veterinary Parasitological Laboratory Techniques*; Ministry of Agriculture 1965).

Flotation method using Sheather's sugar solution

454 g granulated sugar
355 ml warm tap water
6 ml 40% formaldehyde
Dissolve sugar in warm water and add formaldehyde. Follow same method as above, using sugar solution instead of saturated NaCl solution.

Ether sedimentation technique

The ether sedimentation technique (Ritchie 1948; Ridley & Hawgood 1956) is the most satisfactory method for use with primate material. The small samples of sediment obtained from this technique will keep very well for many months if topped up with a few drops of 5% formalin.

(1) Emulsify a sample approximately the size of a pea in 7 ml of 5% formalin in a test tube (Allen & Ridley 1969).
(2) Allow to stand for at least 5 minutes in order to fix the protozoan cysts.
(3) Sieve through a mesh gauze, return to clean tube and make up to 10 ml with 5% formalin.
(4) Add 3 ml of ether, cover the end of the centrifuge tube with a plastic coverslip and shake vigorously until the ether and suspension are thoroughly emulsified.
(5) Centrifuge slowly (1000 rpm) for 2 minutes.
(6) The contents of the centrifuge will have separated into three distinct regions: the upper cm or so of ether, a layer of debris and the formalin suspension medium with some colouring matter.
(7) Loosen the debris at the ether–formalin interface using a swab stick and pour off the contents of the tube in one movement.
(8) The small amount of concentrated deposit is at the bottom of the tube; the drops of liquid on the inside of the glass tube will be sufficient to make smears.
(9) Look at each drop of sediment using iodine solution or Sargeaunt's stain when necessary.

MIF technique (merthiolate–iodine–formaldehyde)

(1) Measure 2.35 ml of MIF stock solution into a Khan tube and insert stopper.
(2) Measure 0.15 ml of Lugol's iodine into a second tube and close.

Pour the MIF solution into the Lugol's solution immediately on collecting the pea-size sample of faecal material and not before. Stir sample and solutions thoroughly. Place stopper in tube. Larger samples may be collected using proportionately larger amounts of Lugol's iodine and MIF solution. To examine sample, draw off supernatant fluid and faeces from the top of the layer, place on a glass slide and examine.

CONCENTRATION METHODS FOR MICROFILARIAE

Nucleopore filtraton technique (Dennis & Kean 1971)

(1) Take the blood sample using heparin or other suitable anticoagulant in the syringe
(2) Using a 5 μ Nucleopore filter, 25 mm in diameter in a Swinnex filter holder, wetting the surface first, pass the blood directly through the filter.
(3) Flush through twice with normal saline (0.85% NaCl) until the washings are clear.
(4) Blow air through the filter using the syringe.
(5) Remove the membrane from the filter blood side upwards and place on a glass slide. (Alternatively, the worms may be fixed by forcing methanol through the filter, blowing dry and then transferring to a slide.)
(6) The membrane can then be stained with haematoxylin or Giemsa as described above.

Millipore membrane technique (Bell 1967; Desowitz & Southgate 1973)

This is basically similar to the Nucleopore technique, using a 25 mm Swinnex filter holder and 5 μ porosity membrane. It differs at (5) where 10 ml of formol saline is passed through the filter, the membrane removed and stained for 30 minutes in a 1:6 Giemsa stain.

MUSCLE DIGESTION TECHNIQUE FOR *TRICHINELLA*

In less severe infections the following muscle digestion technique is required:

(1) Mince the muscle specimen and weigh.
(2) Place the specimen in a 0.5% pepsin and 0.7% hydrochloric acid solution at a temperature of 37 °C, using 10 ml of solution per gram of muscle tissue.
(3) Incubate at 37 °C for 36–48 hours, stirring regularly.
(4) Set up a Baerman apparatus and pour the digest through a double sieve of 40-mesh and 100-mesh per square inch copper gauze into a funnel with a rubber tube clipped at the base.

(5) Rinse the sieve with a little warm water into the funnel.
(6) Leave sieved digest and washings for several hours, during which time the larvae will drop to the bottom of the funnel and can be collected into universal containers.
(7) Pipette onto a slide and examine under a microscope.

REFERENCES

Abaru DE, Denham DA (1976) Laboratory evaluation of a new technique for counting microfilaria in blood. *Trans R Soc Trop Med Hyg* **70**: 333–334

Adams CE, Aitken FC, Worden AN (1967) The rabbit (*Oryctolagus cuniculus*). In: *The UFAW handbook on the care and management of laboratory animals.* 3rd edn (Ed. UFAW). London: Livingstone & UFAW; pp 396–448

Aikawa M (1972) Fine structure of *Haemobartonella* species in the Squirrel Monkey. *J Parasitol* **58**: 628–630

Alger N (1966) A simple, rapid, precise stain for intestinal protozoa. *Am J Clin Pathol* **45**: 361–362

Allen AVH, Ridley DS (1969) Further observations on the formol ether concentration technique for faecal parasites. *J Clin Pathol* **23**: 545–546

Anderson JF, Magnarelli LA, Kutz J (1979) Intra erythrocytic parasites in rodent populations of Connecticut: *Babesia* and *Grahamella* sp. *J Parasitol* **65**: 599–604

Arcay-de-Peraza L (1967) Coccidiosis en monos y su comparación con la isosporosis humana, con descripcion de una nueva especie de *Isospora* en *Cacajao rubicundus* (Uakari monkey o mono chucuto). *Acta Biologica (Venezuelica)* **5**: 203–222

Ash LR (1962) The helminth parasites of rats in Hawaii and the description of *Capillaria tranerae* sp.n. *J Parasitol* **48**: 63–65

Augsten K (1982) Identification of *Eperythrozoon coccoides* by scanning electron microscope. *Z Versuchstierkd* **24**: 169–170

Austwick PKC, Longbottom JL (1980) Medically important *Aspergillus* sp. In: *Manual of clinical microbiology* (Eds. Lenette EH, Balows A, Hauser WJ, Truant J). Washington: American Society for Microbiology; pp 620–627

Bàies A, Suteu I, Klemm W (1968) Notoëdres—Räude des Goldhamsters. *Z Versuchstierkd* **10**: 251–257

Baker EW, Camin JH, Cunliffe F, Wolley TA, Yunker CE (1958) *Guide to the families of mites.* The Institute of Acarology, Dept. of Zoology, University of Maryland, USA

Baker EW, Evans TM, Gould DJ, Hull WB, Keegan HL (1967). *A manual of parasitic mites of medical and economic importance.* New York: Henry Tripp Woodhaven

Baker HJ, Cassell GH, Lindsey JR (1971) Research complications due to *Haemobartonella* and *Eperythrozoon* in experimental animals. *J Pathol* **64**: 625–656

Baker JR (1972) Protozoa of tissues and blood (other than the *Haemosporina*). In: *Pathology of simian primates, Part II* (Ed. Fiennes RNT-W). Basel: Karger; pp 29–56

Balfour A (1922) Observations on wild rats in England, with an account of their ecto and endoparasites. *Parasitology* **14**: 282–298

Barker RJ (1974) Studies in the life cycle and transmission of *Encephalitozoon cuniculi.* PhD Thesis, University of London

Baxby D, Blundell N, Hart CA (1984) The development and performance of a simple, sensitive method for the detection of *Cryptosporidium* oocysts in faeces. *J Hyg (Camb)* **92**: 317–323

Baylis HA, Daubney R (1922) Report on the parasitic nematodes in the collection of the Zoological Survey of India. *Memoirs of the Indian Museum* **7**: 247

Becker ER (1934) *Coccidia and coccidiosis of domesticated, game and laboratory animals and of man.* Ames, Iowa: Collegiate Press

Bedford GAA (1932) A synoptic checklist and host list of the ectoparasites found on South African Mammalia, Aves and Reptilia. *Report on the Division of Veterinary Research in South Africa* **184**: 224–523

Beesley WN (1963). *Cheyletiella parasitivorax* (Acarina: Trombidoida) as a parasitic mite in Britain. *Parasitology* **53**: 651–652

Bell D (1967) Membrane filters and microfilariae: a new diagnostic technique. *Ann Trop Med Parasitol* **61**: 220–223

Bemrick WJ, Erlandsen SL (1988) Giardiasis, is it really a zoonosis? *Parasitology Today* **4**: 69–71

Bennett M, Baxby N, Blundell N, Gaskell CJ, Hart CA, Kelly DF (1985) Cryptosporidiosis in the domestic cat. *Vet Rec* **116**: 73–74

Benoit PLG (1961) Anoploures de Centre African. *Revue de Zoologie et de Botanique Africaines* **63**: 231–241

Beresford-Jones, W. P. (1965) Occurrence of the mite *Psorergates simplex* in mice. *Aust Vet J* **41**: 289–290

Beresford-Jones WP, Fain A, Thoday KL (1976) Further observations relating to *Trixacarus* (*Caviacoptes caviae*; Fain, Lovell & Hyatt 1972)

in guinea pigs (Acarina; Sarcoptidae). *Acta Zool Pathol Vet* **64**: 33–36

Beverley JKA, Watson WA (1961) Ovine abortion and toxoplasmosis in Yorkshire. *Vet Rec* **73**: 6–11

Beverley JKA, Watson WA (1962) Further studies on toxoplasmosis and ovine abortion in Yorkshire. *Vet Rec* **74**: 548–552

Bezubik B, Furmaga S (1960) The parasites in *Macaca cynomologus* L. from Indonesia. *Acta Parasitol Polonica* **8**: 335–344

Blackmore DK & Owen DG (1968) Ectoparasites: The significance in British wild rodents. In: *Diseases in free-living wild animals* (Ed. McDiarmid A). London: Zoological Society & Academic Press; pp 197–220

Blagg W, Schoegel EL, Mansour NS, Khalaf G (1955). A new concentration technique for the demonstration of protozoa and helminth eggs in faeces. *Am J Trop Med Hyg* **4**: 23–28

Boecker H (1953). Die Entwicklung des Kaninchenoxyur en *Passalurus ambiguus. Zeitschrift für Parasitenkunde* **15**: 491–518

Bonciu CB, Clecner A, Greceami A, Margineanu A (1957) Contribution à l'etude de l'infection naturelle des cobayes avec des Sporozoaires au genre 'klossiella'. *Arch Roum Pathol Exp Microbiol* **16**: 131–143

Bonfante R, Faust EC, Giraldo LE (1961) Parasitologic surveys in California. Departmento del valle, Colombia. IX. Endoparasites of rodents and cockroaches in Ward Siloe California, Colombia. *J Parasitol* **47**: 843–846

Bonne C, Sandground JH (1939) On the production of gastric tumors bordering on malignancy, in Javanese monkeys through the agency of *Nochtia nochti* a parasitic nematode. *Am J Cancer* **37**: 173–185

Bornstein S, Iwarsson K (1980) Nasal mites in a colony of Syrian hamsters. *Lab Anim* **14**: 31–33

Bostrom RE, Farrell JF, Martin JE (1968) Simian amoebiasis with lesions simulating human amoebic dystentry. Abstract 51, 19th American Meeting, American Association of Laboratory Animal Science, Las Vegas, USA

Brack M (1987) *Agents transmissible from simians to man*. Berlin: Springer-Verlag

Bray RS, Garnham PCC (1962) The Giemsa colophonium method for staining protozoa in tissue sections. *Ind J Malariol* **16**: 153–155

Bronswijk JEMHV, de Kreek EJ (1976) *Cheyletiella* (Acari: Cheyletiellidea) of dog, cat and domesticated rabbit: A review. *J Med Entomol* **13**: 315–327

Brown RJ, Hinkle DK, Trevethan WP, Kupper JL, McKee AE (1973) Nosematosis in a squirrel monkey (*Saimiri sciureus*). *J Med Primatol* **2**: 114–123

Brug SL, Haga J (1930) Notes on the *Sarcoptes* found in a case of scabies crustosa and in a case of scabies in a monkey. *Mededelingen van den Dienst der Volksgezondheid in Nederlandsch Indische* **19**: 221

Brumpt L, Dechambre E, Desportes C (1939) Prophylaxie et traitement utilisés pour combattre les acanthocéphales parasites des makis de la ménagerie du Jardin des Plantes. *Bulletin de Académie Véterinaire de France* **12**: 198–202

Brumpt E, Joyeaux C (1912) Sur un infusoire nouveau, parasite du chimpanzé *Troglodytella abrassanti* n.g., n.sp. *Bull Soc Pathol Exot Filiales* **5**: 499–503

Brumpt E, Urbain A (1938) Épizootie vermineuse par Acanthocéphales (*Prosthenorchis*) ayant sévi à la singerie du musée de Paris. *Ann Parasitol Hum Comp* **16**: 289–300

Bull PC (1953) Parasites in the wild rabbit *Oryctolagus cuniculus* (L) in New Zealand. *New Zealand Journal of Science and Technology* **Sec. B**: 341–372

Bullock FD, Curtis MR (1924) A study of the reactions of the tissues of the rats liver to the larvae of *Taenia crassicolis* and the cystogenesis of *Cysticercus* sarcoma. *J Cancer Res* **8**: 446–487

Burrows RB (1965) *Microscopic diagnosis of the parasites of man*. New Haven: Yale University Press.

Burrows RB (1967) Improved preparation of polyvinyl alcohol. $HgCl_2$ fixative used for faecal smears. *Stain Technol* **42**: 93–95

Burrows RB (1972) Protozoa of the intestinal tract. In: *Pathology of simian primates, Part II* (Ed. Fiennes RNT-W). Basel: Karger; pp 2–28

Burrows RB (1976) A new fixative and techniques for the diagnosis of intestinal parasites. *Am J Clin Pathol* **48**: 342–346

Burrows RB, Klink GE (1955) *Endamoeba polecki* infections in man. *Am J Hyg* **62**: 156–167

Canavan WPN (1929) Nematode parasites of vertebrates in the Philadelphia zoological garden and vicinity. *Parasitology* **21**: 63–102

Catchpole J, Norton CC (1979) The species of *Eimeria* in rabbits for meat production in Britain. *Parasitology* **79**: 249–257

Cerná Z (1962) Contribution to the knowledge of coccidia parastic in Muridae. *Vestnik Ceskoslovenske Spolecnosti Zoologike* **26**: 1–13

Cerná Z, Senaud J (1969) *Eimeria pragensis* sp. nova. A new coccidian parasite from the intestine of mice. *Folia Parasitologica* **16**: 171–175

Chalupsky J (1973) Detection of *Encephalitozoon cuniculi* in rabbits by the indirect immunofluorescent antibody test. *Folia Parasitol (Praha)* **20**: 281–284

Chalupsky J, Bedrnik P, Vavra J (1971) The indirect fluorescent antibody test for *Nosema cuniculi. J Protozool* **18**: 47 (Abstract 177)

Chan KF (1952) Chemotherapeutic studies on *Syphacia obvelata* infection in mice. *Am J Hyg* **56**: 22–30

Chandler AC (1926) Some parasitic roundworms of the rabbit with descriptions of 2 new species. *Proceedings of the United States National Museum* **66**: 1–6

Chordi A, Walls KW, Kagan IG (1964) Studies on the specificity of the indirect haemagglutination test for toxoplasmosis. *J Immunol* **93**: 1024–1033

Cockrell BY, Valerio MG, Garner FM (1974) Cryptosporidiosis in the intestines of rhesus monkeys (*Macaca mulatta*). *Lab Anim Sci* **24**: 881–887

Coffee GM, McConnell EE (1971) *Pneumonyssus vocalis* n.sp. (Acarina: Alalarachnidae) a new mite from the laryngeal ventricles of the Chacma baboon (*Papio ursinus* (Kerr)). *Zeitschrift für Angewandte Zoologie* **58**: 265–270

Cook R (1956) Murine ear mange: The control of *Psorergates simplex* infestation. *Br Vet J* **112**: 22–25

Cook R (1969) Common diseases of laboratory animals. In: *The IAT manual of laboratory animal practice* (Ed. Short DJ, Woodnott DP). Springfield, Ill: Charles C Thomas; pp 160–215

Cordero del Campillo M (1959) I observaciones sobre el periodo pre-patente, esporulacion morfologia de los ooquistes y estudio biometrico de los mismos, produccion de ooquistes y patogenicidad. *Revista Iberica de Parasitologia* **19**: 351–368

Cosgrove GE, Nelson BM, Gengozian N (1968) Helminth parasites of the tamarin *Saquinus fuscicollis*. *Lab Anim Care* **18**: 654–656

Cosgrove GE, Nelson BM, Self JT (1970) The pathology of pentastomid infection in primates. *Lab Anim Care* **20**: 354–360

Cox FEG (1980) Human babesiosis. *Nature* **285**: 535–536

Cox JC, Pye D (1975) Serodiagnosis of nosematosis using cell culture grown organisms. *Lab Anim* **9**: 297–304

Cram EB (1928) A species of *Bertiella* in man and Chimpanzees in Cuba. *Am J Trop Med Hyg* **8**: 339–344

Curasson G (1929) *Troglodytella alrassanti* infusoire pathogène du chimpanzé. *Ann Parasitol Hum Comp* **7**: 466–468

Current WL, Reese NC, Ernst JV, Bailey WS, Heyman MB, Weinstein WM (1983) Human cryptosporidioisis in immunocompetent and immunodeficient persons. *N Engl J Med* **308**: 1252–1257

da Cunha AM, de Freitas G (1940) Ensaio monografico da familia Cyathodiniidae. *Mem Inst Oswaldo Cruz* **25**: 457–494

da Cunha AM, Muniz J (1929) Nota sobre os parasitas intestinaes do Macacus rhesus com a descripcao de uma nova especie de Octomitus. *Mem Inst Oswaldo Cruz* **5** (Suppl): 34–35

Dalgard D (1980) Note on *Anatrichosoma* in primates and request for information. *Laboratory Primate Newsletter* **19(1)**: 7

Davidson WL, Hummeler K (1960) B virus infection in man. *Ann NY Acad Sci* **85**: 970–979

Davies JHT (1951) Cat itch: *Cheyletiella* and *Notoedres* compared. *Br J Dermatol* **53**: 18–24

Davis JEW (1954) Studies of the sheep mite *Psorergates ovis*. *Am J Vet Res* **15**: 235–257

de Léon DD (1964) Helminth parasites of rats in San Juan Puerto Rico. *J Parasitol* **50**: 478–479

Delorme M (1926) Transmission experimental de *Sarcoptes scabei* var *cuniculi* au cynocéphali (*Papio sphinx*). *Bull Soc Pathol Exot Filiales* **19**: 899

Denham DA (1975) The diagnosis of filariasis. *Ann Soc Belg Med Trop* **55**: 517–524

Dennis DT, Kean BH (1971) Isolation of microfilariae: report of a new method. *J Parasitol* **57**: 1146–1147

Desmonts G, Couvreur J (1974) Toxoplasmosis in pregnancy and its transmission to the foetus. *Bull NY Acad Med* **50**: 146–159

Desmonts G, Forestier F, Thulliez PH, Daffos F, Capella-Pavlovsky M, Chartier M (1985) Prenatal diagnosis of congenital toxoplasmosis. *Lancet* **1**: 500–503

Desowitz RS, Southgate BA (1973) Studies on filariasis in the Pacific. 2. The persistence of microfilaraemia in diethylcarbamazine treated populations of Fiji and Western Samoa: Diagnostic applications of the membrane filtration techniques. *Asian Journal of Tropical Medicine and Public Health* **4**: 179–183

Desportes C (1945) Sur *Strongyloides stercoralis* (Bavay 1876) et sur les *Strongyloides* de primates. *Ann Parasitol Hum Comp* **20**: 160–190

Desportes C, Roth P (1943) Helminths récoltes au course d'autopsies practiquées sur différents mammifères morts à la ménagerie du Musée de Paris. *Bulletin Museum National Histoire Naturel (Paris)* **15**: 108–114

Dick TA, Quentin JC, Freeman RS (1973) Redescription of *Syphacia mesocriceti* (Nematoda: Oxyuroidea) parasite of the golden hamster. *J Parasitol* **59**: 256–259

Doenhoff MJ, Hassounah OA, Lucas SB (1985) Does the immunopathology induced by schistosome eggs potentiate parasite survival? *Immunol Today* **6**: 203–205

Dove WE (1950) The control of laboratory pests and parasites of laboratory animals. In: *The care and breeding of laboratory animals* (Ed. Farris EJ). New York: John Wiley; pp 478–487

Dubey JP (1976) Reshedding of *Toxoplasma* oocysts by chronically infected cats. *Nature* **262**: 213–214

Dubey JP (1978) Life cycle of *Isospora ohioensis* in dogs. *Parasitology* **77**: 1–11

Dubey JP, Mehlhorn H (1978) Extra intestinal stage of *Isospora ohioensis* from dogs in mice. *J Parasitol* **64**: 689–695

Dubey JP, Sharma SP (1980) Parasitaemia and tissue infection in sheep fed *Toxoplasma gondii* cysts. *J Parasitol* **66**: 111–114

Dubey JP, Wong MM (1978) Experimental *Hammondia hammondi* infection in monkeys. *J Parasitol* **64**: 551–552

Duke BOL (1960) Studies on Loiasis in monkeys. III. The pathology of the spleen in drills (*Mandrillus leucophaeus*) infected with Loa loa. *Ann Trop Med Parasitol* **54**: 141–146

Dunn FL (1961) *Molineus vexillarius* sp.n. (Nematoda: Trichostrongylidae) from a Peruvian primate *Tamarinus nigricollis* (Spix 1823). *J Parasitol* **47**: 953–956

Dunn FL (1968) The parasites of *Saimiri* in the context of platyrrhine parasitism. In: *The squirrel monkey* (Eds. Kosenblum LA, Coope RW). New York: Academic Press; pp 31–68

Dunn FL, Greer WE (1962) Nematodes resembling *Ascaris lumbricoides* L.1758 from a Malayan gibbon *Hylobates agilis* Cuvier 1821. *J Parasitol* **48**: 150

Durr U, Pellérdy L (1969) Zum Sauerstoffuerbrauch der Kokzidienoocysten während der Sporulation. *Acta Vet Hung* **19**: 307–310

Dvorak JA, Jones AW, Kuhlman HH (1961) Studies on the biology of *Hymenolepis microstoma* (Dujardin 1845). *J Parasitol* **47**: 833–838

Eaton GJ (1972) Intestinal helminths in inbred strains of mice. *Lab Anim Sci* **22**: 850–853

Eberhard ML (1978) *Tetrapetalonema (T) mystaxi* sp.n. (Nematoda: Filaroidea) from Brazilian moustached marmosets *Saguinus m. mystax*. *J Parasitol* **64**: 204–207

Edgar G (1933) Mortality in rabbits. A note on the occurrence of *Passalurus ambiguus* (Rudolphi) in the Australian rabbit. *Agricultural Gazette of New South Wales* **44**: 470

Eichorn A, Gallagher B (1916) Spontaneous amoebic dysentery in monkeys. *J Infect Dis* **19**: 395–407

Elton C, Ford EB, Baker JR, Gardner AD (1931) Health and parasites of a wild mouse population. *Proc Zool Soc London* **3**: 657–721

English PB (1960) Notoedric mange in cats, with observations on treatment with malathion. *Aust Vet J* **36**: 85–88

Enigk K, Grittner I (1951) Die Sarcoptesräude des Goldhamsters. *Zeitschrift für Parasitenkunde* **15**: 25–33

Ernst JV, Chabotar B, Hammond DM (1971) The oocysts of *Eimeria vermiformis* sp.n. and *E. papillata* sp.n. (Protozoa: Eimeriidae) from the mouse *Mus musculus*. *J Protozoology* **18**: 221–223

Esslinger JH (1962) Hepatic lesions in rats experimentally infected with *Porocephalus crotali* (Pentastomida). *J Parasitol* **48**: 631–638

Esslinger JH (1979) *Tetrapetalonema (T) panamensis* (McCoy 1936) comb. n. (Filaroidea Onchoceridae) in Columbian primates with a description of the adults. *J Parasitol* **65**: 924–927

Estes PC, Richter CB, Franklin JA (1971) Demodectic mange in the golden hamster. *Lab Anim Sci* **21**: 825–828

Ewing HE (1927) Descriptions of three new species of sucking lice together with a key to some related species of the genus *Polyplax*. *Proceedings of the Entomological Society of Washington* **29**: 118–121

Ewing HE (1932) A new sucking louse from the Chimpanzee. *Proceedings of the Biological Society of Washington* **45**: 117–118

Ewing HE (1933) The taxonomy of the Anopleura genus *Pediculus* Linnaeus. *Proceedings of the Biological Society of Washington* **46**: 167–174

Ewing HE (1938) North American mites of the subfamily Myobiinae new subfamily (Arachnida). *Proceedings of the Entomological Society of Washington* **40**: 180–197

Fain A (1958) Un nouveau parasite de l'orang utan. *Rhinophaga pongicola* n.sp. (Acarina: Halarachnidae). *Revue de Zoologie et de Botanique Africaines* **58**: 323–327

Fain A (1959*a*) Deux nouveau genres d'acariens vivant dans l'epaissent des muqueses nasale et buccale chez un lemurien (Trombidiformes, Demodicidae). *Bulletin de l'Academie de Société Royale de Belgique* **95**: 263–273

Fain A (1959*b*) Les Acariens du genre *Pneumonyssus* des singes au Congo Belge (Halarachnidae: Mesostigmata). *Ann Parasitol Hum Comp* **34**: 126–148

Fain A (1960) La pentastome chez l'homme. *Bulletin de l'Académie Royale de Médecine de Belgique*, Series VI, **25**: 16–32

Fain A (1961) Les pentastomides de l'Afrique Central. Musée Royal de l'Afrique Central Annales Serie 8°. *Sciences Zoologiques* **92**: 1–115

Fain A (1966) Pentastomids of snakes, their parasitological role in man and animals. *Memorias do Instituto Butantan* (Sao Paulo) **33**: 167–174

Fain A (1968) Étude de la variabilite de *Sarcoptes scabiei* avec une revision des Sarcoptidae. *Acta Zoologica et Pathologica Antverpiensia* **47**: 3–196

Fain A, Lukoschus F (1968) Note sur deux acariens parasites nasicoles de murides (Ereynetidae: Trombidiformes). *Bulletin et Annales de la Société d'Entomologie de Belgique* **104**: 85–90

Fain A, Hovell GJR, Hyatt KH (1972). A new sarcoptid mite producing mange in albino guinea-pigs. *Acta Zoologica et Pathologica Antverpiensia* **56**: 73–82

Farenholtz H (1910*a*) Neue Läuse. *Jahresbericht des Niedersächsischen Zoologischen Vereius Hannover* **1**: 57–75

Farenholtz H (1910*b*) Diagnosen neuer Anopleuren. *Zoologische Anzeiger* **35**: 714–715

Farenholtz H (1916) Weitere Beitrage zur Kenntris der Anopleuren. *Archiv für Naturgeschichte A* **11**: 1–34

Faubert GM (1988) Is giardiasis a true zoonosis? Evidence that giardiasis is a zoonosis. *Parasitology Today* **4**: 66–68

Faust EC (1933) Experimental studies on human and primate species of *Strongyloides*. II. The development of *Strongyloides* in the experimental host. *Am J Hyg* **18**: 114–132

Faust EC, Beaver PC, Jung RC (1968) *Animal agents and vectors of human disease*. 3rd edn. Philadelphia: Lea and Febiger

Ferris GF (1916) A catalogue and host list of the Anopleura. *Proceedings of the California Academy of Science* (4th series) **6**: 129–213

Ferris GF (1932) *Contributions towards a monograph of the sucking lice, Part V*. California: Stanford University Publications

Ferris GF (1951) The sucking lice. *Pacific Coast Entomological Society Memoirs* **1**: 1–320

Fiennes RNT-W (1967) *Zoonoses of primates. The epidemiology and ecology of simian diseases in relation to man*. New York: Cornell University Press

Fiennes RNT-W (1972) Ectoparasites and vectors. In: *Pathology of simian primates, Part II* (Ed. Fiennes RNT-W). Basel: Karger; pp 158–173

Filice FP (1952) Study on the cytology and life history of a *Giardia* from the laboratory rat. *University of California Publications in Zoology* **52**: 53–145

Finkeldy W (1931) Pathologisch-anatomische Befunde bei der oesophagostomiasis der Javeneraffen. *Zeitschrift für Infektionskrankheiten, Parasitäre Krankheiten und Hygiene der Haustiere* **40**: 146–164

Fisher RL (1963) *Capillaria hepatica* from the rock vole in New York. *J Parasitol* **49**: 450

Fleck DG, Chessum BS, Perkins M (1972) Coccidian-like nature of *Toxoplasma gondii*. *Br Med J* **3**: 111–112

Flynn RJ (1955) Ectoparasites of mice. *Proceedings of the Animal Care Panel* **6**: 75–91

Flynn RJ (1973) *Parasites of laboratory animals*. Ames: Iowa State University Press

Foster HL (1963) Specific pathogen free animals. In: *Animals for research: principles of breeding and management* (Ed. Lane-Petter W). New York: Academic Press; pp 110–138

Foster AC, Johnson CM (1939) A preliminary note on the identity, life cycle and pathogenicity of an important nematode parasite of captive monkeys. *Am J Trop Med* **19**: 265–277

Fox H (1927) Notes on special animals. *Report of Laboratory and Museum of Comparative Pathology, Philadelphia*; pp 12–22

Fox H (1930) Animals of special importance. *Report of Laboratory and Museum of Comparative Pathology, Philadelphia*; pp 11–18

Frenkel JK (1977) *Besnoitia wallecei* of cats and rodents with a reclassification of other cyst forming Isosporid coccidia. *J Parasitol* **63**: 611–628

Frenkel JK, Dubey JP (1972) Rodents as transport hosts for cat coccidia *Isospora felis* and *I. rivolta*. *J Infect Dis* **125**: 69–72

Frenkel JK, Dubey JP (1975) *Hammondia hammondi* gen nov sp.nov. from domestic cats, a new coccidian related to *Toxoplasma* and *Sarcocystis*. *Zeitschrift für Parasitenkunde* **46**: 3–12

Fulton JD, Joyner LP (1948) Natural amoebic infections in laboratory rodents. *Nature* **161**: 66–68

Gannon J (1978) The immunoperoxidase test diagnosis of *Encephalitozoon cuniculi* in rabbits. *Lab Anim* **12**: 125–127

Gannon J (1979) *Encephalitozoon cuniculi*: a serological and histological survey of laboratory rodents and lagomorphs with observations on pathology and transmission in rabbits and mice. MPhil Thesis, University of London

Garcia LS, Bruckner DA, Brewer TC, Schimizu RY (1983) Techniques for recovery and identification of *Cryptosporidium* oocysts from stool specimens. *J Clin Microbiol* **18**: 185–189

Gard S (1945) A note on the coccidium *Klossiella muris*. *Acta Pathol Microbiol Scand* **22**: 427–434

Garnham PCC (1966) *Malaria parasites and other haemosporidia*. Oxford: Blackwell Scientific

Garnham PCC, Heisch RB, Minter DM (1961) The vector of *Hepatocystis* (*Plasmodium*) *kochi*: the successful conclusion of observations in many parts of tropical Africa. *Trans R Soc Trop Med Hyg* **55**: 497–502

Garnham PCC, Pick F (1952) Unusual form of merocysts of *Hepatocystis* (*Plasmodium*) *kochi*. *Trans R Soc Trop Med Hyg* **46**: 535–537

Gebauer O (1933) Bertrag zur kenntris von Nematoden aus Affenlungen. *Zeitschrift für Parasitenkunde* **5**: 724–734

Geck P (1971) India ink immunoreaction for the rapid detection of enteric pathogens. *Acta Microbiol Acad Sci Hung* **18**: 191–196

Gieman QM (1935) Cytological studies of the *Chilomastix* (Protozoa: Flagellata) of man and other mammals. *J Morphol* **57**: 429–459

Gleason NN, Healy CR, Western KA, Benson GD, Schultz MG (1970) The 'Grey' strain of *Babesia microti* from a human case established in laboratory animals. *J Parasitol* **56**: 1256–1257

Gledhill AW (1956) Quantitative aspects of the enhancing action of *Eperythrozoa* on the pathogenicity of mouse hepatitis virus. *J Gen Microbiol* **15**: 292–303

Goodman DG, Garner FM (1972) A comparison of methods for detecting *Nosema cuniculi* in rabbit urine. *Lab Anim Sci* **22**: 568–572

Goodpasture FB (1924) Spontaneous encephalitis in rabbits. *J Infect Dis* **34**: 428–432

Gordon RM & Lavoirpierre MMJ (1962) *Entomology for students of medicine*. Oxford: Blackwell Scientific

Gordon RM, Unsworth K, Seaton DR (1943) The development and transmission of scabies as studies in rodent infections. *Ann Trop Med Parasitol* **37**: 174–194

Gordon RM, Webber WA (1954) A new technique for the concentration of microfilaria from the venus blood and application to their detection in persons harbouring them in low density together with the observations of significance of such low densities. *Ann Trop Med Parasitol* **49**: 80–95

Graham GL (1960) Parasitism in monkeys. *Ann NY Acad Sci* **85**: 842–860
Gunther C (1942) Notes on the Listrophoridae (Acarina: Sarcoptoidea). *Proceedings of the Linnean Society of New South Wales* **67**: 109–110
Haberkorn A (1970) Die Entwicklung von *Eimeria falciformis* (Eimer 1870) in der weissen maus (*Mus musculus*). *Zeitschrift für Parasitenkunde* **34**: 49–69
Habermann RT, Williams FP (1958) The identification and control of helminths in laboratory animals. *J Natl Cancer Inst* **20**: 979–1010
Hanson RP, Sulkin SE, Buescher EL, Hammon W, Hammon McD, McKinney RW, Work TH (1967) Arbovirus infections of laboratory workers. *Science* **158**: 1283–1286
Hashimoto I, Honjo S (1966) Survey of helminth parasites in cynomologus (sic) monkeys (*Macaca irus*). *Jpn J Med Sci Biol* **19**: 218
Hegner RW (1923) Giardias from wild rats and mice and *Giardia caviae* sp.n. from the guinea-pig. *Am J Hyg* **3**: 345–349
Hegner RW (1934) Intestinal protozoa of chimpanzees. *Am J Hyg* **19**: 480–501
Hegner RW, Chu HJ (1930) A comparative study of the intestinal protozoa of wild monkeys and man. *Am J Hyg* **12**: 62–108
Hegner RW, Rees CW (1933) *Taliferria clarki*, a new genus and species of ciliate from the caecum of the red spider monkey *Ateles geoffroyi* Kuhl. *Transactions of the American Microbiological Society* **52**: 317–321
Hendricks LD (1974) A redescription of *Isospora arctopitheci* Rodhain 1933 (Protozoa Eimeriidae) from primates of Panama. *Proceedings of the Helminthological Society of Washington* **41**: 229–233
Henriksen SA, Pohlenz JFL (1981) Staining of cryptosporidia by a modified Ziehl-Neelsen technique. *Acta Vet Scand* **22**: 594–596
Henry DP (1932) Coccidiosis of the guinea-pig. *University of California Publications in Zoology* **37**: 211–268
Henschele WP (1961) Internal parasitism of monkeys with the Pentastomid *Armillifer armillatus*. *J Am Vet Med Assoc* **139**: 911–912
Heyneman DJ (1953) Autoinfection of white mice resulting from infection by *H. nana*. *J Parasitol* **39**(Suppl): 28 (Abstract)
Heyneman D (1961) Studies on helminth immunity: III. Experimental verification of autoinfection from cysticercoids of *Hymenolepis nana* in the white mouse (Cestoda: Hymenolepididae). *J Infect Dis* **109**: 10–18
Hill WCO, Porter A, Southwick MD (1952) The natural history, endoparasites and pseudoparasites of the tarsiers (*Tarsius carbonarius*) recently living in the Society's menagerie. *Proc Zool Soc London* **122**: 79–119
Hirst S (1915) Preliminary list of the Acari occurring on the brown rat (*Rattus norvegicus*) in Great Britain and the description of a new species (*Haemogamasus oudemonsi*). *Bulletin of Entomological Research* **5**: 119–124
Hirst S (1917) Remarks on a certain species of the genus *Demodex* Owen (the *Demodex* of man, the horse, dog, rat and mouse). *Annals of the Magazine of Natural History London*, Series 8, **20**: 232–235
Hirst S (1919) *Studies on the Acari. I: The genus Demodex (Owen).* London: British Museum
Hirst S (1922) *Mites injurious to domestic animals* (British Museum (Natural History) Economic Series No. 13). London: British Museum
Honigberg BM, Balamuth W, Bovee ED, *et al.* (1964) A revised classification of the Phylum Protozoa. *J Protozool* **11**: 7–20
Hopkins GHE (1949–50) Host associations of the lice of mammals. *Proc Zool Soc London* **119**: 387–604
Hsu C-K (1982) Protozoa. In: *The Mouse in Biomedical Research, Vol. 2: Diseases* (Eds. Foster ML, Small JD, Fox JG). London & New York: Academic Press; pp 359–372
Hsu C-K, Melby EC (1974) *Isospora callimico* n.sp. (Coccidia: Eimeriidae) from Goeldi's marmoset (*Callimico goeldii*). *Lab Anim Sci* **24**: 476–479
Hull WB (1970) Respiratory mite parasites in non human primates. *Lab Anim Care* **20**: 402–406
Hunninen AV (1935*a*) A method of demonstrating Cysticercoids of *Hymenolepis fraterna* (*H. nana var fraterna*) in the intestinal villi of mice. *J Parasitol* **21**: 124–125
Hunninen AV (1935*b*) Studies in the life history and host parasitism of *H. fraterna* (*H. nana var fraterna*) in white mice. *Am J Hyg* **22**: 414–443
Hunninen AV (1936) An experimental study of internal autoinfection with *Hymenolepis fraterna* in white mice. *J Parasitol* **22**: 84–87
Hussey KL (1957) *Syphacia muris* v.s. *S. obvelata* in laboratory rats and mice. *J Parasitol* **43**: 555–559
Inglis WG (1961) The oxyurid parasites (Nematoda) of primates. *Proc Zool Soc London* **136**: 103–122
Innes JRM (1969) Pulmonary acariasis: An enzootic disease caused by *Pneumonyssus simicola* in simians, particularly *Macaca mulatta*. In: *Laboratory Animal Handbooks 4: Hazards of handling simians* (Eds. Perkins FT, O'Donoghue PN). London: Laboratory Animals Ltd; pp 101–105
Innes JRM, Hull WB (1972) Endoparasites—lung mites. In: *Pathology of simian primates, Part II* (Ed. Fiennes RNT-W). Basel: Karger; pp 177–193
Innes JRM, Leman W, Frenkel JK, Bonners G (1962) Occult endemic encephalitozoonosis of the central nervous system of mice (Swiss Bagg O'Grady strain). *J Neuropathol Exp Neurol* **21**: 519–533
Irvin AD, Young ER, Osborn GD, Francis LMA (1981) A comparison of *Babesia* infections in intact, surgically splenectomised and congenitally asplenic (Dh/ +) mice. *Int J Parasitol* **11**: 251–255

Jackson RF (1969) Diagnosis of heartworm disease by examination of the blood. *J Am Vet Med Assoc* **154**: 374

Jacobson RH, Reed ND (1974) The thymus dependency of resistance to pinworm infection in mice. *J Parasitol* **60**: 976–979

Jervis HR, Merril TG, Sprinz H (1966) Coccidoisis in the guinea-pig small intestine due to a *Cryptosporidium*. *Am J Vet Res* **27**: 408–414

Johnson CM (1941) Observations on natural infections of *Entamoeba histolytica* in water at temperatures between 45 °C and 55 °C. *Am J Trop Med* **30**: 53–58

Johnson PT (1958) Type specimens of lice (Order: Anopleura) in the United States National Museum. *Proceedings of the United States National Museum* **108**: 39–49

Jones TC (1967) Pathology of the liver of rats and mice. In: *The Pathology of Laboratory Rats and Mice* (Eds. Cotchin E, Roe FJC). Oxford: Blackwell Scientific; pp 1–23

Joseph SA (1970) On the fleas of the genus Ctenocephalides. *Madras Veterinary College Annual* **28**: 12–13

Joyeaux C, Baer JG (1945) Morphologie, évolution et position systematique de *Cataenotaenia pusilla* (Goeze 1782) parasite de rongeurs. *Revue Suisse de Zoologie* **52**: 13–51

Joyeaux CE, Foley H (1930) Les helminthe de *Meriones shawi* Rezet dans le nord de L'Algerie. *Bulletin de la Société Zoologique de France*, **55**: 353–374

Joyeaux CE, Kobozieff NI (1928) Recherches sur *l'Hymenolepis microstoma* (Dujardin 1845). *Ann Parasitol Hum Comp* **6**: 59–79

Jurumilinta R, Maegraith BG (1961) Intestinal amoebiasis associated with amoebic liver abscesses produced by intracaecal inoculation of *Entamoeba histolytica* in hamsters. *Ann Trop Med Parasitol* **55**: 383–388

Kagan IG (1974) Advances in the immunodiagnosis of parasitic infections. *Zeitschrift für Parasitenkunde* **45**: 163–195

Karr SL, Henrickson RV, Else JG (1979) A survey for *Anatrichosoma* (Nematoda: Trichinellida) in wild caught *Macaca mulatta*. *Lab Anim Sci* **29**: 789–790

Karr SL, Wong MM (1975) A survey of *Sarcocystis* in non human primates. *Lab Anim Sci* **25**: 641–645

Kassel R, Levitan S (1953) A jugular technique for the repeated bleeding of small mammals. *Science* **118**: 563–564

Keast D, Chesterman FC (1972) Changes in macrophage metabolism in mice heavily infected with *Hexamita muris*. *Lab Anim* **6**: 33–39

Kellett BS, Bywater JEC (1978) A modified india-ink immuno-reaction for the detection of encephalitozoonosis. *Lab Anim* **12**: 59–60

Kessel JF, Johnstone HG (1949) The occurrence of *Entamoeba polecki* (Prowazek 1912) in *Macaca mulatta* and in man. *Am J Trop Med* **29**: 311–317

Kheisin EM (1938) Observations on the morphology of *Entamoeba coli* var *cuniculi* Brug from the rabbit intestine (in Russian). *Vestnik Microbiologii i Epidemiologie i Parazitologie* **17**: 384–390

Kim KC, Emerson KC (1968) Descriptions of two species of pediculidae (Anopleura) from great apes (Primates: Pongidae). *J Parasitol* **54**: 690–695

Knezevich AL, McNulty WP jr (1970) Pulmonary acariasis (*Pneumonyssus simicola*) in colony bred *Macaca mulatta*. *Lab Anim Care* **20**: 693–696

Knott J (1939) A method for making microfilarial surveys on dry blood. *Trans R Soc Trop Med Hyg* **68**: 177–186

Koller LD (1969) Spontaneous *Nosema cuniculi* infection in laboratory rabbits. *J Am Vet Med Assoc* **155**: 1108–1114

Kovatch RM, White JD (1972) Cryptosporidiosis in two juvenile rhesus monkeys. *Vet Pathol* **9**: 426–440

Kraus AL (1974) Arthropod parasites. In: *The biology of the laboratory rabbit* (Eds. Weisbroth SA, Flatt RE, Kraus AL). New York: Academic Press; pp 287–316

Kreier JP, Ristic M (1968) Haemobartonellosis, eperythrozoonosis, grahamellosis and ehrlichosis. In: *Infectious blood diseases of man and animals, Vol II* (Eds. Weinman D, Ristic M). New York: Academic Press; pp 387–472

Kreier JP, Ristic M (1974) Haemobartonella. In: *Manual of determinative bacteriology*. 8th edn (Eds. Buchanan RE, Gibbons NE). Baltimore: Williams & Wilkins; Part 18. Family III Genus IV, pp 910–912

Kunstyr I (1977) Infectious form of *Spironucleus* (*Hexamita*) *muris*: banded cysts. *Lab Anim* **11**: 185–188

Kuntz RE (1972) Trematodes of the intestinal tract and biliary passages. In: *Pathology of simian primates, Part II* (Ed. Fiennes RNT-W). Basel: Karger; pp 104–121

Kuntz RE, Myers BJ (1967) Primate cysticercosis *Taenia hydatigena* in Kenya vervets (*Cercopithecus aethiops* Linnaeus 1758) and Taiwan macaques (*Macaca cyclopis* Swinhoe 1864). *Primates* **8**: 83–88

Kuntz RE, Myers BJ, Bergner J, Armstrong JF (1968) Parasites and commensals of the Taiwan macaque (*Macaca cyclopis* Swinhoe 1862). *Formosan Science* **22**: 120–136

Kuntz RE, Myers BJ, Katzberg A (1970) Spargenosis and 'proliferative like' spargena in vervets and baboons from East Africa. *J Parasitol* **56**: 196–197

Kuntz RE, Myers BJ, Vice TE (1967*a*) Intestinal protozoans and parasites of the gelada baboon. (*Theropithecus gelada* Rüppel 1835). *Proceedings of the Helminthological Society of Washington* **34**: 65–66

Kuntz RE, Myers BJ, Vice TE (1967*b*) Microbiological parameters of the baboon (*Papio* sp.): Parasitology. In: *The baboon in medical research, Vol 2* (Ed. Vagtborg H). Austin: University of Texas Press; pp 741–755

Kutzer E, Grünberg W (1967) Prosarcoptesträude (*Prosarcoptes pitheci* Philippe 1948) bei einen kapuzineraffen (Cebus c. capuchinus). *Zeitschrift für Parasitenkunde* **28**: 290–298

Laing ABG, Edeson JFB, Wharton RH (1960) Studies on filariasis in Malaya; the vertebrate hosts of *Brugia malayi* and *B. pahangi*. *Ann Trop Med Parasitol* **54**: 92–99

Lainson R, Garnham PCC, Killick-Kenrick R, Bird RG (1964) Nosematosis, a microsporidial infection of rodents and other mammals including man. *Br Med J* **2**: 470–472

Lapage G (1940) The study of coccidiosis (*Eimeria caviae* [Sheather 1924]) in the guinea-pig. *Veterinary Journal* **96**: 144–154, 190–202, 242–254, 280–295

Lapage G (1968) *Veterinary parasitology*. London: Oliver & Boyde

Lavoirpierre MMJ (1960) Notes acarologiques 2: Quelques remarques sur *Trixacarus diversus* Sellnick 1966 (= *Sarcoptes anaranthes* Guilnon 1946) et sur trois espèces récemment décrites de Sarcoptes de singes it des Chauve-Souris. *Ann Parasitol Hum Comp* **35**: 166–170

Lavoirpierre MMJ (1970) A note on the sarcoptid mites of primates. *J Med Entomol* **7**: 376–380

Lavoirpierre MMJ, Crew W (1955) The occurrence of a mange mite *Psorergates* sp. (Acarina) in a West African monkey. *Ann Trop Med Parasitol* **49**: 351

Lawless DK, Burrows RG (1965) *Microscopic diagnosis of the parasites of man, Vol 1*. Yale University Press

Layne JN (1967) Incidence of *Porocephalus crotali* (Pentastomida) in Florida mammals. *Bulletin of the Wildlife Diseases Association* **3**: 105–109

Lebel RR, Nutting WB (1973) Demodectic mites of subhuman primates I: *Demodex saimiri* sp.n. (Acari: Demodicidae) from the squirrel monkey, *Saimiri sciureus*. *J Parasitol* **59**: 719–722

Leerhoy J, Jensen HS (1967) Sarcoptic mange in a shipment of cynomolgus monkeys. *Nordisk Veterinärmedicin* **19**: 128–130

Levine ND (1961) *Protozoan parasites of laboratory animals and of man*. Minneapolis: Burgess

Levine ND (1968) *Nematode parasites of domestic animals and of man*. Minneapolis: Burgess

Levine ND (1973) *Protozoan parasites of domestic animals and of man*. 2nd edn. Minneapolis: Burgess

Levine ND (1984) Taxonomy and review of the coccidian genus *Cryptosporidium* (Protozoa, Apicomplexa). *J Protozool* **31**: 94–98

Levine ND, Ivens V (1965) *The coccidian parasites (Protozoa: Sporozoa) of rodents. III* (Biological Monographs of the University of Illinois Press). Urbana: University of Illinois Press

Levine ND, Tadros W (1980) Named species and hosts of *Sarcocystis* (Protozoa: Apicomplexa: Sarcocystidae). *Systemic Parasitology* **2**: 41–59

Lewis WP, Kessel JF (1961) Haemagglutination in the diagnosis of toxoplasmosis and amoebiasis. *Arch Ophthalmol* **66**: 471–476

Lindsay DS, Stuart BP, Wheat BE, Ernst JV (1980) Endogenous development of the swine coccidium *Isospora suis* Biester 1934. *J Parasitol* **66**: 771–779

Litchford RG (1963) Observations on *Hymenolepis microstoma* in three laboratory hosts: *Mesocricetus auratus*, *Mus musculus* and *Rattus norvegicus*. *J Parasitol* **49**: 403–410

Little MD (1966) Comparative morphology of six species of *Strongyloides* (Nematoda) and redefinition of the genus. *J Parasitol* **52**: 69–84

Lucas MS (1932*a*) A study of *Cyathodinium piriforme*, an endozoic protozoan from the intestinal tract of the guinea-pig. *Archiv für Protistenkunde* **77**: 64–72

Lucas MS (1932*b*) The cytoplasmic phases of rejuvenescence and fission in *Cyathodinium piriforme*. II. A type of fission heretofore undescribed for ciliates. *Archiv für Protistenkunde* **77**: 407–423

Lussier G, Marois P (1964) Animal sarcosporidiosis in the province of Quebec. *Can J Public Health* **55**: 243–246

McCarrison R (1920) The effects of deficient diet on monkeys. *Br Med J* **1**: 249–253

McConnell EE, Basson PA, de Vos V (1972) Laryngeal acariasis in the Chacma baboon. *J Am Vet Med Assoc* **161**: 678–682

McConnell EE, de Vos AJ, Basson PA (1971) *Isospora papionis* n.sp. (Eimerida) of the Chacma baboon *Papio ursinus* (Kess, 1792). *J Protozool* **18**: 28–32

Machado-Filho DA (1950) Revisào do gènero *Prosthenorchis* Travassos 1915 (Acanthocephala). *Mem Inst Oswaldo Cruz* **48**: 495–544

Maede Y (1975) Studies on feline haemobartonellosis. IV. Lifespan of erythrocytes of cats infected with *Haemobartonella felis*. *Jpn J Vet Sci* **37**: 269–272

Maede Y, Hata R (1975) Studies on feline haemobartonellosis. II. The mechanism of anaemia produced with *Haemobartonella felis*. *Jpn J Vet Sci* **37**: 49–54

Maede Y, Sonada M (1975) Studies on feline haemobartonellosis. III. Scanning electron microscopy of *Haemobartonella felis*. *Jpn J Vet Sci* **37**: 209–211

Mandour AM (1964) A report on the presence of *Sarcocystis* in different hosts. *Trans R Soc Trop Med Hyg* **58**: 287–288

Mandour AM (1969) *Sarcocystis nesbitti* n.sp. from the rhesus monkey. *J Protozool* **16**: 353–354

Marcus MB, Killick-Kendrick R, Garnham PCC (1974) The coccidial nature and life cycle of *Sarcocystis*. *J Trop Med Hyg* **77**: 248–259

Margulis L, Corliss JO, Melkonian M, Chapman DJ (1989) *Handbook of the protista*. Boston: Jones & Bartlett

Martinez-Palermo A (1987) The pathogenesis of amoebiasis. *Parasitology Today* **3**: 111–118

Matheson R (1950) *Medical entomology*. 2nd edn. Ithaca, NY: Comstock

Matubayasi H (1938) Studies on parasitic protozoa in Japan: IV. Coccidia parasitic in wild rats (*Epimys rattus alexandrinus* and *E. norvegicus*). *Annotationes Zoologicae Japonaises* **17**: 144-163

Mead-Briggs AR (1964*a*) Some experiments concerning the interchange of rabbit fleas *Spilopsyllus cuniculi* (Dale) between living rabbit hosts. *J Anim Ecol* **33**: 13-26

Mead-Briggs AR (1964*b*) The reproductive biology of the rabbit flea *Spilopsyllus cuniculi* (Dale) and the dependence of this species upon the breeding of its host. *J Exp Biol* **41**: 371-402

Mead-Briggs AR (1965) Reproduction in the European rabbit flea and its dependence upon the reproductive state of the host. In: *Proceedings of the XII International Congress of Entomology*. London; pp 802-803

Mead-Briggs AR, Vaughan JA, Dennison BD (1975) Seasonal variation in number of the rabbit flea on the wild rabbit. *Parasitology* **70**: 103-118

Meshorer A (1969) Hexamitiasis in laboratory mice. *Lab Anim Care* **19**: 33-38

Miller JH (1959) *Hepatocystis* (= *Plasmodium*) *kochi* in the dog-face baboon *Papio doguera*. *J Parasitol* **45** (Suppl): 53 (Abstract)

Miller MJ (1947) Studies on the life history of *Trichocephalus vulpis*, the whipworm of dogs. *Can J Vet Res* **25**: 1-12

Miller MJ (1952) The experimental infection of *Macaca mulatta* with human strains of *Entamoeba histolytica*. *Am J Trop Med Hyg* **1**: 417-428

Miller MJ, Bray RS (1966) *Entamoeba histolytica* infections in the Chimpanzee (*Pan satyrus*). *J Parasitol* **52**: 386-388

Ministry of Agriculture, Fisheries and Food (1965) *Manual of Veterinary Parasitological Laboratory Techniques* (Technical Bulletin No. 18). London: HMSO

Mirelman D (1987*a*) Ameba-bacterium relationship in amoebiasis. *Microbiol Rev* **51**: 272-284

Mirelman D (1987*b*) Effect of culture conditions and bacterial associates on the zymodeme of *Entamoeba histolytica*. *Parasitology Today* **3**: 37-40

Mirelman D, Bracha R, Chayen A (1986) *Entamoeba histolytica*: Effect of growth conditions and bacterial associates on isoenzyme patterns and virulence. *Exp Parasitol* **62**: 142-148

Moore JA, Kuntz E (1981) *Babesia microti* infections in non-human primates. *J Parasitol* **67**: 454-455

Moreton HL (1969) Spargenosis in African green monkeys (*Cercopithecus aethiops*). *Lab Anim Care* **19**: 253-255

Mougeot G, Golvan YJ, Delattne P (1978) Detection of schistosomiasis in rats using micro-blood sampling and immunofluorescence during an epidemiological study. *Comptes Rendus Hebdomadaires des Séances de l'Académie des Sciences* **286**: 141-143

Myers BJ, Kuntz RE (1965) A check-list of parasites reported for the baboon. *Primates* **6**: 137-194

Myers BJ, Kuntz RE (1967) Parasites of baboons taken by the Cambridge Mwanza Expedition (Tanzania, 1965). *East Afr Med J* **44**: 322-324

Neal RA (1969) Intestinal protozoa of simians as a handling hazard. In: *Laboratory Animal Handbooks 4: Hazards of handling simians* (Eds. Perkins FT, O'Donoghue PN). London: Laboratory Animals Ltd; pp 77-81

Neal RA, Harris WG (1975) Attempt to infect inbred strains of rats and mice with *Entamoeba histolytica*. *Trans R Soc Trop Med Hyg* **69**: 29-30

Neal RA, Harris WG (1977) Attempts to infect inbred strains of rats and mice with *Entamoeba histolytica*. *Protozoology* **3**: 197-199

Neal RA, Vincent P (1955) Strain variation in *Entamoeba histolytica*. I. Correlation of invasiveness in rats with the clinical history and treatment of experimental infections. *Parasitology* **45**: 152

Neal RA, Vincent P (1956) Strain variation in *Entamoeba histolytica*. II. The effect of serial liver passages on the virulence. *Parasitology* **46**: 173-182

Nelson B, Cosgrove GE, Gengozian N (1966) Diseases of an imported primate *Tamarinus nigricollis*. *Lab Anim Care* **16**: 255-275

Nie D (1950) Morphology and taxonomy of the intestinal protozoa of the guinea pig *Cavia porcella*. *J Morphol* **86**: 381-493

Nuttall GHF (1919) The sytematic position synonymy and iconography of *Pediculus humanus* and *Phthirus pubis*. *Parasitology* **11**: 329-346

Nutting WB (1961) *Demodex aurati* sp. nov. and *D. criceti*, ectoparasites of the golden hamster (*Mesocricetus auratus*). *Parasitology* **51**: 515-522

Nutting WB (1965) Host parasite relations. Demodicidae. *Acaralogica* **7**: 301-317

Oldham JN (1967) Helminths, ectoparasites and protozoa in rats and mice. In: *Pathology of laboratory rats and mice* (Eds. Cotchin E, Roe FJC). Oxford: Blackwell Scientific; pp 641-679

Ollett W (1951) Further observations on the Gram-Twort stain. *J Pathol Bacteriol* **63**: 166

Olsen OW (1967) *Animal parasites: Their biology and life cycles*. 2nd edn. Minneapolis: Burgess

Orihel TC (1968) Development of *Necator americanus* and *Ascaris lumbricoides* from man and lower primates. In: *Proceedings of the 8th International Congress for Tropical Medicine and Malaria*. Teheran; p 174 (Abstract)

Orihel TC (1972) Anatrichosomiasis in African monkeys. *J Parasitol* **56**: 982-985

Orihel TC, Seibold HR (1972) Nematodes of the bowel and tissues. In: *Pathology of simian primates, Part II* (Ed. Fiennes RNT-W). Basel: Karger; pp 76-103

Ott KJ, Stauber LA (1967) *Eperythrozoon coccoides*: Influence on the course of infection of *Plasmodium chabaudi* in the mouse. *Science* **155**: 1546-1548

Owen DG (1974) A study of the life cycles of *Eimeria stiedae* (Lindemann 1865) and *Eimeria falciformis* (Eimer, 1870) in specific pathogen free and gnotobiotic animals. PhD Thesis, University of Liverpool

Owen DG (1982) *Haemobartonella muris*: case report and investigation of transplacental transmission. *Lab Anim* **16**: 17–19

Owen DG (1985) A mouse model for *Entamoeba histolytica* infection. *Lab Anim* **19**: 297–304

Owen DG (1987) *Entamoeba histolytica* in inbred mice: evidence of liver involvement. *Trans R Soc Trop Med Hyg* **81**: 617–620

Owen DG, Gannon J (1980) Investigation into the transplacental transmission of *Encephalitozoon cuniculi* in rabbits. *Lab Anim* **14**: 35–38

Owen DG, Turton J (1979) Eradication of the pinworm *Syphacia obvelata* from an animal unit by anthelmintic therapy. *Lab Anim* **13**: 115–118

Owen DG, Young C (1973) The occurrence of *Demodex aurati* and *D. criceti* in the United Kingdom. *Vet Rec* **92**: 282–284

Parke HK (1961) *Toxoplasma* haemagglutination test. *Arch Ophthalmol* **65**: 184–191

Patten RA (1939) Amoebic dysentery in orang-utans (*Simia satyrus*). *Aust Vet J* **15**: 68–71

Pattison M, Clegg FG, Duncan AL (1971) An outbreak of encephalomyelitis in broiler rabbits caused by *Nosema cuniculi*. *Vet Rec* **88**: 404–405

Pellérdy LP (1965) *Coccidia and coccidiosis*. Budapest: Akadémiai Kiado Publishing House of the Hungarian Academy of Sciences

Pellérdy LP, Durr U (1970) Zum endogenen Entwicklungszyklus von *Eimeria stiedae* (Lindemann 1865) Lissbalt & Hartmann. *Acta Vet Acad Sci Hung* **20**: 221–224

Peters W (1965) Competitive relationship between *Eperythrozoon coccoides* and *Plasmodium berghei* in the mouse. *Exp Parasitol* **16**: 158–166

Petri M (1969) Studies on *Nosema cuniculi* found in transplantable tumours with a survey of microsporidiosis in mammals. *Acta Pathol Microbiol Scand*; Suppl 204

Philippe J (1948) Note sur les gales du singe. *Bull Soc Pathol Exot Filiales* **41**: 597

Phillipson RF (1974) Intermittent egg release by *Aspiculuris tetraptera* in mice. *Parasitology* **69**: 207–213

Poelma FG (1966) *Eimeria lemuris* n.sp., *E. galago*, n.sp. and *E. orolicni* n.sp. from a galago *Galago senegalensis*. *J Protozool* **13**: 547–549

Poindexter HA (1942) A study of the intestinal parasites of the monkeys of the Santiago Island primate colony. *Puerto Rico Journal of Public Health and Tropical Medicine* **18**: 175–191

Poole TB, ed (1972) *The UFAW handbook on the care and management of laboratory animals*. 6th edn. Edinburgh & London: Churchill Livingstone, for Universities Federation for Animal Welfare

Porter A (1966) *Chilomastix tarsii* n.sp. a new flagellate from the gut of *Tarsius carbonarius*. *Proc Zool Soc London* **121**: 915

Porter DA, Otto GF (1934) The guinea pig nematode *Paraspidodera uncinata*. *J Parasitol* **20**: 323

Porter JA jr (1972) Parasites of marmosets. *Lab Anim Sci* **22**: 503–506

Powell EC, McCarley JB (1975) A murine *Sarcocystis* that causes an *Isospora*-like infection in cats. *J Parasitol* **61**: 928–931

Pryor WH, Bradbury RP (1975) *Haemobartonella canis* infection in research dogs. *Lab Anim Sci* **25**: 566–569

Rahaman H (1975) Ectoparasitism in *Macaca radiata*. *Lab Anim Care* **25**: 505

Raulston GL (1972) Psorergatic mites in Patas monkeys. *Lab Anim Sci* **22**: 107–108

Reardon LV, Rininger BF (1968) A survey of parasites in laboratory primates. *Lab Anim Care* **18**: 577–580

Reichenow E (1920) Den Wiederkauer-Infusoren rervandte Formen aus Gorilla und Schimpanse. *Archiv für Protistenkunde* **41**: 1–33

Reichenow E (1953) *Lehrbuch der Protozoenkunde, vol II*. Jena: Gustav Fischer

Richards OW, Davies RG, eds (1977) *Imms' general textbook of entomology, 10th edn. Vol 2: Classification and biology*. London: Chapman & Hall

Ridley DS, Hawgood BS (1956) The value of formol ether concentration of faecal cysts and ova. *J Clin Pathol* **9**: 74–76

Rijpstra AC (1967) Sporocysts of *Isospora* sp. in a chimpanzee (*Pan troglodytes*). *Proceedings Koninkliijke Nederlandse Academie van Wetenschappen* **70**: 395–401

Ritchie LS (1948) An ether sedimentation technique for routine stool examination. *Bulletin of the US Army Medical Department* **8**: 326

Robinson JJ (1954) Common infectious disease of laboratory rabbits questionably attributed to *Encephalitozoon cuniculi*. *Arch Pathol* **58**: 71–84

Roth EF, Tanowitz H, Wittner M, Yoshihiro U, Hsieh HS, Neumann G, Nagel RL (1981) *Babesia microti*: Biochemistry and function of hamster erythrocytes infected from human sources. *Exp Parasitol* **51**: 116–123

Rothschild M, Ford B (1973) Factors influencing the breeding of the rabbit flea (*Spillopsyllus cuniculi*). A springtime accelerator and a Kairomone in nestling rabbit urine with notes on *Cedlopsylla simplex* another hormone bound species. *J Zool* **170**: 87

Roudabush RL (1937) The endogenous phases of the life cycle of *Eimeria nieschulzi*, *Eimeria separata* and *Eimeria miyairii*, coccidian parasites of the rat. *Iowa State College Journal of Science* **11**: 135–163

Rowland E, Vandenbergh JG (1965) A survey of intestinal parasites in a new colony of rhesus monkeys. *J Parasitol* **51**: 294–295

Ruch TC (1959) *Diseases of laboratory primates*. Philadelphia: WB Saunders

Rudnick P, Hollingsworth JW (1959) Lifespan of rat erythrocytes parasitised by *Bartonella muris*. *J Infect Dis* **104**: 24–27

Ruebush TK, Cassaday PB, Marsh HJ, Lisker SA, Voorhees DB, Mahoney EA, Healy GR (1977) Human babesiosis on Nantucket Island. *Ann Intern Med* **86**: 6–9

Ruebush TK, Collins WE, Healy GR, Wamen MCW (1979) Experimental *Babesia microti* infection in non splenectomised *Macaca mulatta*. *J Parasitol* **65**: 144–146

Ruiz A, Frenkel JK (1976) Recognition of cyclic transmission of *Sarcocystis muris* by cats. *J Infect Dis* **133**: 409–418

Rutherford RL (1943) The life cycle of four intestinal coccidia of the domestic rabbit. *J Parasitol* **29**: 10–32

Sabin AB, Feldman HA (1948) Dyes as microchemical indicators of a new immunity phenomenon affecting a protozoan parasite (*Toxoplasma*). *Science* **108**: 660–663

Sandground JH (1931) Studies on the life history of *Ternidens diminutus*, a nematode parasite of man, with observations on its occurrence in certain regions of S. Africa. *Ann Trop Med Parasitol* **25**: 147–180

Sapero JJ, Lawless DK (1953) The MIF stain preservation technique for the identification of intestinal protozoa. *Am J Trop Med Hyg* **2**: 613–619

Sargeaunt PG (1962) Confirmation of amoebic cyst chromatoids by wet stain after Ridleys faecal concentration. *Trans R Soc Trop Med Hyg* **56**: 12

Sargeaunt PG (1987) The reliability of *Entamoeba histolytica* zymodemes in clinical diagnosis. *Parasitology Today* **3**: 40–43

Sargeaunt PG, Williams JE (1978) Electrophoretic isoenzyme patterns of *Entamoeba histolytica* and *Entamoeba coli*. *Trans R Soc Trop Med Hyg* **72**: 164–166

Sargeaunt PG, Williams JE (1979) Electrophoretic isoenzyme patterns of the pathogenic and non-pathogenic intestinal amoebae of man. *Trans R Soc Trop Med Hyg* **73**: 225–227

Schacher JF, Khalil GM, Salman S (1965) A field study of *halzoun* (parasitic pharyngitis) in Lebanon. *J Trop Med Hyg* **68**: 226–230

Schullhorn van Veen TW, Blotkamp J (1978) Histochemical differentiation of microfilariae of *Dipetalonema*, *Dirofilaria*, *Onchocerca* and *Setaria* sp. of man and domestic animals in the Zaria Area (Nigeria). *Tropenmedizin und Parasitologie* **27**: 33–35

Schulz RE, Krepkorgorskaja TA (1932) *Dentostomella translucida* n.gen n.sp (Nematoda: Oxyuridae) aus einem Nagetier (*Rhomomys opimus* Licert). *Zoologische Anzeiger* **97**: 330–334

Schwarzbrott SS, Wagner JE, Frisk C (1974) Demodicosis in the mongolian gerbil (*Meriones unguiculatus*): a case report. *Lab Anim Sci* **24**: 666–668

Seamer J, Gledhill AW, Barlow JL, Hotchin J (1961) The effect of *Eperythrozoon coccoides* upon lymphocytic choriomeningitis in mice. *J Immunol* **86**: 512–515

Sebesteny A (1969) Pathogenicity of intestinal flagellates in mice. *Lab Anim* **3**: 71–77

Sebesteny A (1974) The transmission of intestinal flagellates between mice and rats. *Lab Anim* **8**: 74–81

Sebesteny A (1979) Transmission of *Spironucleus* and *Giardia* spp. and some non pathogenic intestinal protozoa from infested hamsters to mice. *Lab Anim* **13**: 189–191

Self JT, Cosgrove GE (1968) Pentastome larvae in laboratory primates. *J Parasitol* **54**: 969

Self JT, Cosgrove GE (1972) Pentastomida. In: *Pathology of simian primates, Part II* (Ed. Fiennes RNT-W). Basel: Karger; pp 194–204

Self JT, McMurry FB (1948) *Porocephalus crotali* Humbolt (Pentastomida) in Oklahoma. *J Parasitol* **34**: 21–23

Shadduck JA, Pakes SP (1978) Protozoal and metazoal diseases. In: *Pathology of Laboratory Animals*, Vol II (Eds. Benirschke K, Garner FM, Jones TC). New York: Springer Verlag; chapter 17

Sheldon WG (1966) Psorergatic mange in the sooty mangabey (*Cercocebus torquates atys*) monkey. *Lab Anim Care* **16**: 276–279

Sholten T (1972) An improved technique for the recovery of intestinal protozoa. *J Parasitol* **58**: 633–634

Shortt HE, and Blackie EJ (1965) An account of the genus *Babesia* found in certain small mammals in Britain. *J Trop Med Hyg* **68**: 37–42

Siebold HR, Fussel EN (1973) Intestinal microsporidiosis in *Callicebus moloch*. *Lab Anim Sci* **23**: 115–118

Simpson CF, Gaskin JM, Harvey JW (1978) Ultrastructure of erythrocytes parasitised by *Haemobartonella felis*. *J Parasitol* **64**: 504–511

Small E, Ristic M (1967) Morphologic features of *Haemobartonella felis*. *Am J Vet Res* **28**: 845–851

Smetana HF, Orihel TC (1969) Gastric papillomata in *Macaca speciosa* induced by *Nochtia nochti* (Nematoda Trichostrongylidae). *J Parasitol* **55**: 349–351

Smiley RL (1965) A new mite *Prosarcoptes scanloni* from monkey (Acarina: Sarcoptidae). *Proceedings of the Entomological Society of Washington* **67**: 166–167

Smit FGA (1957*a*) Siphonaptera. In: *Handbook for the identification of British insects, Vol. 1 Part 16*. London: Royal Entomological Society; pp 1–94

Smit FGA (1957*b*) The recorded distribution and hosts of Siphonaptera in Britain. *Entomologists Gazette* **8**: 45–75

Smith WN, Chitwood MB (1967) *Trichospirura leptostoma*, gen. et sp.n., from pancreatic ducts of white-eared marmosets. *J Parasitol* **53**: 1270–1272

Smith WN, Levy S (1969) The effects of *Trichospirura leptostoma* on the pancreas of *Callithrix jacchus*. 44th Annual Meeting of the American Society of Pathology, Washington; Abstract No. 70, p 45

Smith CEG, Simpson DIH, Zlotnik IO (1967) Fatal human disease from vervet monkeys. *Lancet* **2**: 1119–1121

Soulsby EJL (1965) *Textbook of veterinary clinical parasitology, Vol 1: Helminths.* Philadelphia: FA Davis

Stephenson AC (1915) *Klossiella muris. Quarterly Journal of Microscopical Science* **61**: 127–135

Stiles CW, Hassall A (1929) A key catalogue of parasites reported for primates (monkeys & lemurs) with their possible public health importance and key catalogue of primates for which parasites are reported. *Bulletin 152, Hygiene Laboratories United States Public Health Service*; pp 409–601

Stunkard HW (1965) New intermediate hosts in the life cycle of *Prosthenorchis elegans* (Diesing 1851) an Acanthocephalan parasite of primates. *J Parasitol* **51**: 645–649

Suldey EW (1924) Dysenteric amebienne spontanée chez le chimpanzé (*Troglodytes niger*). *Bull Soc Pathol Exot Filiales* **17**: 771–773

Suzuki K, Suto T, Fujita J (1965) Serological diagnosis of toxoplasmosis by indirect immunofluorescent staining. *Bulletin of the National Institute of Health (Tokyo)* **5**: 73–85

Sweatman GK (1958) On the life history and validity of the species in *Psoroptes*. A genus of mange mites. *Can J Zool* **36**: 905–929

Swezey WW (1932) The transition of *Troglodytella abrassarti* and *Troglodytella abrassarti uminata*, intestinal ciliates of the chimpanzee, from one type to another. *J Parasitol* **19**: 12–16

Swift HF, Boots RH, Miller CP (1922) A cutaneous nematode infection in monkeys. *J Exp Med* **35**: 599–620

Tanaka HM, Fukui M, Yamamoto H, Hayam S, Kodera S (1962) Studies on the identification of common intestinal parasites of primates. *Bull Exp Anim* **11**: 111–116

Thienpont D, Mortelmans SJ, Vercruysse J (1962) Contribution a l'étude de la Trichurlose du chimpanzé et de son traitement avec la Methyridine. *Ann Soc Belg Med Trop* **2**: 211–218

Todd KS, Lepp DL (1971) The life cycle of *Eimeria vermiformis*. Ernst. Chobotar and Hammond, 1971 in the mouse *Mus musculus. J Protozool* **18**: 332–337

Toft JD II, Schmidt RE, De Paoli A (1976) Intestinal polyposis associated with oxyurid parasites in a chimpanzee (*Pan troglodytes*). *J Med Primatol* **5**: 360–364

Townsend GH (1969) The grading of commercially bred laboratory animals. *Vet Rec* **83**: 225–226

Travassos LP, Vogelsang EG (1929) Sobre um novo Trichostrongylidae parasito de *Macacus rhesus. Sciencia Medica (Rio de Janeiro)* **7**: 509–511

Twort JM, Twort CC (1923) Disease in relation to carcinogenic agents among 60,000 experimental mice. *J Pathol Bacteriol* **35**: 219–242

Tyzzer EE (1907) A sporozoan found in the peptic glands of the common mouse. *Proc Soc Exp Biol Med* **5**: 12–13

Tyzzer EE (1910) An extracellular coccidium *Cryptosporidium muris* (gen. et. sp. nov.) on the gastric glands of the common mouse. *Archiv für Protistenkunde* **26**: 394–412

Tyzzer EE (1912) *Cryptosporidium parvum* (sp.nov.) a coccidium found in the small intestine of the common mouse. *Archiv für Protistenkunde* **26**: 394–412

Tzipori S, Smith M, Halpin C, Angus KW, Sherwood D, Campbell I (1983) Experimental cryptosporidiosis in calves: Clinical manifestations and pathological findings. *Vet Rec* **112**: 116–120

Upton SJ, Current WL (1985) The species of *Crypto sporidium* (Apicomplexa: Cryptosporidiidae) infecting mammals. *J Parasitol* **71**: 625–629

Van Riper OC, Day JW, Fineg J, Prine JR (1966) Intestinal parasites of recently imported chimpanzees. *Lab Anim Care* **16**: 360–363

Vetterling JM, Takeuchi A, Madden PA (1971) Ultra structure of *Cryptosporidium wrairi* from the guinea pig. *J Protozool* **18**: 248–260

Vierra LQ, Moraes-Santos T (1987) *Schistosoma mansoni*: evidence for a milder response in germfree mice. *Rev Inst Med Trop Sao Paulo* **29**: 37–42

Vogue M, Hyneman D (1957) Development of *Hymenolepis nana* and *Hymenolepis diminuta* (Cestoda Hymenolepidae) in the intermediate host *Tribolium confusum. University of California Publications in Zoology* **59**: 549–580

von Sternberg C (1929) *Klossiella muris*—ein haufiger, anscheinend wenigbekannter Parasit der weissen maus. *Wien Klin Wochenschr* **42**: 419–421

Waller T (1975) Growth of *N. cuniculi* in established cell lines. *Lab Anim* **9**: 61–68

Waller T (1977) The india-ink immunoreaction: a method for the rapid diagnosis of encephalitozoonosis. *Lab Anim* **11**: 93–97

Wallberg JA, Stark DM, Desch C, McBride DF (1981) Demodicosis in laboratory rats (*Rattus norvegicus*). *Lab Anim Sci* **31**: 60–62

Wantland WW (1955) Parasitic fauna of the golden hamster. *J Dent Res* **34**: 631–649

Watson DP (1961) The effect of the mite *Myocoptes musculinus* (C. L. Koch 1840) on the skin of the white laboratory mouse and its control. *Parasitology* **51**: 373–378

Watson WA, Beverley JKA (1971) Epizootics of toxoplasmosis causing ovine abortion. *Vet Rec* **83**: 120–123

Webber WAF (1955*a*) The filarial parasites of primates, a review. I. *Dirofilaria* and *Dipetalonema. Parasitology* **49**: 123–141

Webber WAF (1955*b*) The filarial parasites of

primates, a review. II. *Loa. Protofilaria* and *Parlitimosa* with notes on incompletely identified adult larval forms. *Parasitology* **49**: 235–249

Weisbroth SH, Scher S (1969) *Trichosomoides crassicauda* infection of a commercial rat breeding colony. I. Observations on life cycle and propagation. 20th Annual Meeting of the American Association of Laboratory Animal Science, Dallas; Abstract 40

Wenrich DH (1933) A species of *Hexamita* (Protozoa, Flagellata) from the intestine of a monkey (*Macacus rhesus*). *J Parasitol* **19**: 225–229

Wenrich DH (1937) Studies on *Iodamoeba buetschlii* (Protozoa) with special reference to nuclear structure. *Proceedings of the American Philosophical Society* **77**: 183–205

Wenrich DH (1947) The species of *Trichomonas* in man. *J Parasitol* **33**: 177–188

Wenrich DH, Nie D (1949) The morphology of *Trichomonas wenyoni* (Protozoa: Mastigophora). *J Morphol* **85**: 519–531

Wenrich DH, Saxe LH (1950) *Trichomonas microti* n.sp. (Protozoa: Mastigophora). *J Parasitol* **36**: 261–269

Wenyon CM (1926) *Protozoology* (2 vols). London: Wood

Wightman SR, Pilitt PA, Wagner JE (1978*a*) *Dentostomella translucida* in the mongolian gerbil (*Meriones unguiculatus*). *Lab Anim Sci* **28**: 290–296

Wightman SR, Wagner JR, Corwin RM (1978*b*) *Syphacia obvelata* in the mongolian gerbil (*Meriones unguiculatus*). Natural occurrence and experimental transmission. *Lab Anim Sci* **28**: 51–54

Williams GAH (1959) Experimental hepatic amoebiasis and its application to chemotherapeutic studies. *Br J Pharmacol Chemotherapy* **14**: 488–492

Wilson JM (1979) The biology of *Encephalitozoon cuniculi*. Review article. *Medical Biology* **57**: 84–101

Wilson JM (1986) Can *Encephalitozoon cuniculi* cross the placenta? *Res Vet Sci* **40**: 138

Witenberg GG (1964) Acanthocephala infections. In: *Zoonosis* (Ed. Van der Hoeden J). New York: Elsevier; pp 708–709

Wong MM, Kozek WJ (1974) Spontaneous toxoplasmosis in macaques: A report of four cases. *Lab Anim Sci* **24**: 273–278

Worms MJ (1967) Parasites in newly imported animals. *Journal of the Institute of Animal Technology* **18**: 39–47

Wright JH, Craighead EM (1922) Infectious motor paralysis in young rabbits. *J Exp Med* **36**: 135–140

Wright VS (1985) A comparison of the development of *Eimeria nieschulzi* (Dieben, 1924) and *Eimeria miyarii* (Ohira, 1912) in the small intestine of the rat. Dissertation submitted for Master of Science (Laboratory Animal Sciences), University of London

York W, Maplestone PA (1933) *The nematode parasites of vertebrates*. London & New York: Hafner

Yost DH (1958) *Encephalitozoon* infection in laboratory animals. *J Natl Cancer Inst* **20**: 957–963

Zumpt F (1961) The arthropod parasites of vertebrates in Africa South of the Sahara (Ethiopian region) Vol. I: Chelicerata. *Publications of the South African Institute for Medical Research* **9**: 1–457

Zumpt F, Till W (1954) The lung and nasal mites of the genus *Pneumonyssus* Banks (Acarina: Laelaptidae) with description of two new species from African primates. *Journal of the Entomological Society of Southern Africa* **17**: 195–211

Zumpt F, Till W (1955) The mange causing mites of the genus *Psorergates* (Acarina Myobiidae) with description of new species from a South African monkey. *Parasitology* **45**: 269–274

FURTHER READING

Anderson RC (1971) Lungworms. In: *Parasitic diseases of wild mammals. Part 2: Endoparasites* (Eds. Davis JW, Anderson RC). Iowa State University Press

Chandler AC, Read CP (1961) *Introduction to parasitology*. New York & London: John Wiley

Gershwin ME, Merchant B, eds (1981) *Immunologic defects in laboratory animals*. New York & London: Plenum Press

Heston WE (1941) Parasites. In: *Biology of the laboratory mouse* (Ed. Snell GD). Philadelphia: Blakiston; pp 349–379

Owen D, ed (1981) *Animal models in parasitology* (Symposium held at the Royal Zoological Society, London). London: Macmillan Press

Strandtmann RW, Wharton GW (1958) *Manual of mesostigmated mites*. Institute of Acarology, Department of Zoology, University of Maryland

Weinman D, Ristic M, eds (1968) *Infectious blood diseases of man and animals*. New York: Academic Press

Wenrich DH (1949) Protozoan parasites of the rat. In: *The rat in laboratory investigations* (Eds. Farris EJ, Griffith JQ). Philadelphia: Lippincott; pp 486–501

GLOSSARY

Acanthella Larval stage of Acanthocephala ('thorny headed worms'), found in the body cavity of certain Arthropoda, usually cockroaches and beetles.

Aedeagus A portion of the male genitalia of some insects.

Alae The thin-walled extensions of the body margins found in the anterior or posterior regions of certain Nematoda.

Amastigote The intracellular, round-bodied stage of *Leishmania* which has no free flagellum.

Axostyle A clear axial rod or filament, which provides a flexible support to the bodies of certain Protozoa.

Blepharoplast The basal body of flagellated Protozoa.

Bosses Broad blunt protuberances of the cuticle in helminths.

Bothria Suckers of the scolex of some tapeworms which aid attachment to the host.

Bradyzoite A slowly multiplying zoite of *Toxoplasma gondii* during the chronic stages of infection.

Bursa An expansion of the cuticle, supported by rays, around the male genitalia of some nematode worms.

Caruncle Warty outgrowth.

Cercaria The larval stage of a fluke which develops within a snail host. When released the larva has an oval or heart shaped anterior portion and a tail and is an active swimmer.

Chelicera A prehensile appendage situated close to the mouthparts of certain arthropods such as mites and ticks.

Chitin Complex polysaccharide forming the major part of the horny exoskeleton on Insecta and Crustacea.

Cilia Fine cytoplasmic extensions of an animal cell, which move in a regular rhythm. They provide the motile organelles of certain microorganisms and are present as ciliated epithelium in all Metazoa except the Arthropoda and Nematoda.

Coenurus The larval stage of a metacestode which consists of a large fluid-filled bladder which usually has multiple scolices protruding into it.

Colerette An extension of the cuticle reflected forward and forming a hood-like structure around the mouthparts of some spiruroid nematodes.

Commensal A non-parasitic organism which lives in, on or with another organism and shares the same food.

Contractile vacuole A vesicle in the cytoplasm of certain protozoa, which regularly expands and contracts, expelling to the exterior the excess water which enters the cell from the environment by osmosis.

Coracidium First larval stage of a diphyllobothriid tapeworm. The onchosphere is enclosed in a ciliated coat and swims actively in the water before being eaten by a crustacean.

Costa A ridge or rib shape as found in certain Protozoa where it constitutes a supporting ridge at the base of an undulating membrane.

Coxa The large first joint of the limbs in arthropods.

Ctenidia Rows of spines which form distinctive structures known as combs in insects. A useful characteristic in the identification of the flea.

Cysticanth Mature (infective) acanthella stage of Acanthocephala.

Cysticercoid The larval form of some tapeworms in which the scolex is contained in a small bladder and the body remains solid.

Cysticercus The larval form of some tapeworms in which the scolex is invaginated into a large bladder.

Cytopyge The excretory pore of certain Protozoa.

Cytostome The mouth opening of certain Protozoa.

Ecdysis The shedding of the outer cuticle.

Embryonation Very early development of an organism from a fertilized ovum.

Embryophore A membrane surrounding the developing embryo (onchosphere) in tapeworm eggs.

Empodium A median lobe or spine-like process arising between the claws of some Insecta and Acarina.

Endodyogeny A method of reproduction in certain Protozoa by a process of budding where two daughter cells develop in a mother cell before destroying it.

Endopolygeny A method of protozoan reproduction similar to endodyogeny but differing in that more than two offspring are found within the mother cell and there are more nuclear divisions before merozoite formation.

Epistome Part of an insect head.
Fibril A fine fibre.
Filariform Term applied to a nematode larva which has a long cylindrical oesophagus without a clearly marked posterior bulb.
Flagellum Elongated motile process of a cell or protozoan, which produces currents in fluid media and locomotion of the individual.
Fusiform Spindle-shaped.
Gametocyte Cell which forms a gamete.
Gametogony Process of fusion of two reproductive cells to produce a zygote.
Gamont A reproductive cell which can fuse with the nucleus and cytoplasm of another to produce a new individual.
Gubernaculum Cord supporting testis.
Haemocoele A reduced body cavity found in Arthropoda, which acts as a blood vascular system.
Hydatid Cysticercus stage of *Echinococcus granulosus* which consists of a fluid-filled sac containing numerous larval scolices and daughter cysts.
Hypostome Part of the mouthparts of a tick, usually heavily armed with hooks and used to anchor the organism while feeding.
Karyosome Minute spheres found in the nucleus and said to contain chromatin.
Kinetoplast An organelle present in some Mastigophora and consisting of the blepharoplast and parabasal body and believed to coordinate movement.
Leptomonad A form of trypanosome, long and slender in shape with a free flagellum.
Macrogametocyte The female sexual stage of sporozoan parasites.
Macrogamont Largest gamete in a heterogametic organism.
Macronucleus The larger of two nuclei in ciliated Protozoa, which disappears during conjugation and is reconstituted from the micronucleus.
Maxilla An upper jaw, also an appendage, on the head or cephalothorax of arthropods. Can be modified in different groups.
Merocercoid Incomplete type of pleurocercoid.
Meront A single-celled stage arising from schizogony in the asexual stages of the sporozoan life cycle.
Merozoite A product of the process of schizogony in the life of sporozoan parasites.
Mesothorax Middle of the three segments making up the thorax of an insect.
Metacercaria Larval stage of some trematodes. The cercaria emerging from the snail intermediate host encyst on the surface of vegetation.
Microgametocyte The male sexual stage in the life cycle of a sporozoan parasite.
Micronucleus A small nucleus found in ciliated Protozoa, which divides mitotically and provides the gametes during conjugation.
Micropyle Pore in the egg membrane of an ovum through which the sperm enters during fertilization.
Miracidium First larval stage in the trematode life cycle. The leaf-like organism is ciliated and is released from the egg in water where it swims away in search of a suitable snail host.
Nymph The young form of mites and insects, resembling the adult but lacking sexual organs.
Ocelli Simple light receptors found in many invertebrates.
Onchosphere An embryo tapeworm within the egg.
Operculum A lid at the anterior portion of a louse egg, often with pores or tubercles, which the larva pushes off when ready to emerge.
Palp Part of the first and second maxillae.
Papilla Small protuberance.
Parthenogenetic Organism capable of reproduction without fertilization.
Pedicel A short stalk-like appendage which bears a sucker in some mites.
Pedipalp A small leg-like appendage close to the mouthparts in arachnids.
Pellicle A firm but flexible thin outer covering, as found in some Protozoa.
Peristome A depression of the body wall lined with cilia and leading to the cytostome in certain Protozoa.
Pleurocercoid A larva of certain cestodes having a solid body, particularly those found in fish muscle.
Podosoma Part of the body (idiosoma) of a mite which bears the four pairs of legs.
Prepatent period The period of time between infection of the host with certain Protozoa or helminths and the appearance of eggs or cysts in the faeces.
Procercoid First larval stage of some cestodes.
Proglottid A segment of an adult tapeworm containing at first one or more complete sets of male and female reproductive organs and later, when mature, the eggs.
Promastigote An extracellular stage of *Leishmania* sp., also called a leptomonad form. It is an actively motile stage with a single flagellum.
Prothorax Most anterior of three segments which make up the thorax of an insect.
Pseudopodium Protoplasmic process extruded by certain Protozoa to produce movement or ingest food.
Residual body Waste products of sporulation enclosed within the oocyst or sporocyst.
Rhabditiform Term applied to a nematode larva which has a short straight oesophagus and double bulb consisting of an anterior thick portion and strong posterior bulb connected by a narrow neck region.
Rostellum The anterior extremity of the scolex of a tapeworm, which may be retractile and may also bear hooks.

Scolex The head region and the usual point of attachment of the tapeworm to its host. It may bear grooves, suckers or hooks to aid attachment.
Schizont An asexual dividing stage in the life cycle of certain Protozoa.
Seta A chitinous bristle on the epidermis of an invertebrate.
Sparganum Generic larval stages of certain pseudophyllidean tapeworms found in the muscle and connective tissue of their intermediate host. Now given as a common name of any unidentifiable larval stage found in that situation.
Spermatheca A pouch or sac found in the female or hermaphrodite animal for the reception and storage of spermatozoa.
Spicule A fine cuticular pointed body forming part of the male copulatory apparatus in certain Nematoda. One or two spicules may be present and their size and shape are of great value in the identification of the species.
Sporoblast Cell from which the sporozoite stage develops.
Sporocyst Chitinous envelope containing the sporozoites in the mature oocysts of some Sporozoa.
Sporont An individual which gives rise to a sporozoite generation.
Stichocyte cells Chain of cylindrical cells through which runs the narrow oesophagus in nematode worms of the superfamily Trichinelloidea.
Strobila The ribbon of proglottids which form the body of an adult tapeworm apart from the scolex.
Stylet A slender hollow tube which is part of the piercing mouthparts of some acarines.
Tarsus Region of an insect leg below the tibia which bears terminal claws and setae.
Trophozoite A vegetative stage in the reproduction of Protozoa.
Undulating membrane An aid to locomotion in some Mastigophora, which consists of a delicate membrane extending fully or partially along the length of the body and which has a flagellum running along its outer margin.
Vitelline gland Organ which produces yolk.
Zoite Organism in the cystic form in species of Sporozoa such as *Toxoplasma* and *Sarcocystis*.
Zymodeme Strains of *Entamoeba histolytica* identifiable by their isoenzyme patterns.

INDEX

Page numbers in *italic* indicate the host-parasite lists found in Appendix A1–A5. Page numbers in **bold** indicate an illustration.